Ultrafiltration-Based Extraction for Biological Agents in Early Warning Systems

About the Awwa Research Foundation

The Awwa Research Foundation (AwwaRF) is a member-supported, international, nonprofit organization that sponsors research to enable water utilities, public health agencies, and other professionals to provide safe and affordable drinking water to consumers.

The Foundation's mission is to advance the science of water to improve the quality of life. To achieve this mission, the Foundation sponsors studies on all aspects of drinking water, including supply and resources, treatment, monitoring and analysis, distribution, management, and health effects. Funding for research is provided primarily by subscription payments from approximately 1,000 utilities, consulting firms, and manufacturers in North America and abroad. Additional funding comes from collaborative partnerships with other national and international organizations, allowing for resources to be leveraged, expertise to be shared, and broad-based knowledge to be developed and disseminated. Government funding serves as a third source of research dollars.

From its headquarters in Denver, Colorado, the Foundation's staff directs and supports the efforts of more than 800 volunteers who serve on the board of trustees and various committees. These volunteers represent many facets of the water industry, and contribute their expertise to select and monitor research studies that benefit the entire drinking water community.

The results of research are disseminated through a number of channels, including reports, the Web site, conferences, and periodicals.

For subscribers, the Foundation serves as a cooperative program in which water suppliers unite to pool their resources. By applying Foundation research findings, these water suppliers can save substantial costs and stay on the leading edge of drinking water science and technology. Since its inception, AwwaRF has supplied the water community with more than $300 million in applied research.

More information about the Foundation and how to become a subscriber is available on the Web at **www.awwarf.org**.

Ultrafiltration-Based Extraction for Biological Agents in Early Warning Systems

Prepared by:
Kevin Oshima and **Geof Smith**
New Mexico State University
Department of Biology
Department 3AF
Las Cruces, NM 88003

Jointly sponsored by:
Awwa Research Foundation
6666 West Quincy Avenue, Denver, CO 80235-3098

and

U.S. Department of the Interior
Bureau of Reclamation
Technical Service Center
Water Resources Research Laboratory
Denver, CO 80225

Published by:

Awwa Research Foundation

DISCLAIMER

This study was jointly funded by the Awwa Research Foundation (AwwaRF) and the U.S. Bureau of Reclamation (Reclamation) under Cooperative Agreement No. 00FC810253. AwwaRF and Reclamation assume no responsibility for the content of the research study reported in this publication or for the opinions or statements of fact expressed in the report. The mention of trade names for commercial products does not represent or imply the approval or endorsement of either AwwaRF or Reclamation. This report is presented solely for informational purposes.

CONTENTS

TABLES

FIGURES

FOREWORD

The Awwa Research Foundation (AwwaRF) is a nonprofit corporation that is dedicated to the implementation of a research effort to help utilities respond to regulatory requirements and traditional high-priority concerns of the industry. The research agenda is developed through a process of consultation with subscribers and drinking water professionals. Under the umbrella of a Strategic Research Plan, the Research Advisory Council prioritizes the suggested projects based upon current and future needs, applicability, and past work; the recommendations are forwarded to the Board of Trustees for final selection. The foundation also sponsors research projects through an unsolicited proposal process; the Collaborative Research, Research Applications, and Tailored Collaboration programs; and various joint research efforts with organizations such as the U.S. Environmental Protection Agency, the U.S. Bureau of Reclamation, and the Association of California Water Agencies.

This publication is a result of one of these sponsored studies, and it is hoped that its findings will be applied in communities throughout the world. The following report serves not only as a means of communicating the results of the water industry's centralized research program but also as a tool to enlist the further support of the nonmember utilities and individuals.

Projects are managed closely from their inception to the final report by the foundation's staff and large cadre of volunteers who willingly contribute their time and expertise. The foundation serves a planning and management function and awards contracts to other institutions such as water utilities, universities, and engineering firms. The funding for this research effort comes primarily from the Subscription Program, through which water utilities subscribe to the research program and make an annual payment proportionate to the volume of water they deliver and consultants and manufacturers subscribe based on their annual billings. The program offers a cost-effective and fair method for funding research in the public interest.

A broad spectrum of water supply issues is addressed by the foundation's research agenda: resources, treatment and operations, distribution and storage, water quality and analysis, toxicology, economics, and management. The ultimate purpose of the coordinated effort is to assist water suppliers to provide the highest possible quality of water economically and reliably. The true benefits are realized when the results are implemented at the utility level. The foundation's trustees are pleased to offer this publication as a contribution toward that end.

Walter J. Bishop
Chair, Board of Trustees
Awwa Research Foundation

Robert C. Renner, P.E.
Executive Director
Awwa Research Foundation

ACKNOWLEDGMENTS

The authors of this report are indebted to the following individuals at New Mexico State University for their cooperation and participation in this project: David Acosta, Marija Borjan, Dabanjana Dasgupta, Laura Grainger, Laura Jones, and Sal Lopez.

The assistance of the Awwa Research Foundation project manager, Misha Hasan, is appreciated. The following Project Advisory Committee members are also greatly appreciated for their input to this project: Thomas Gargan, Janet Jenson, Dennis Juranek, Mark LeChevallier, Alan Lindquist, and Jeremy Walker.

EXECUTIVE SUMMARY

This project developed methods for the rapid and simultaneous extraction of a wide variety of pathogens from drinking water when confronted with the potential for intentional introduction of a biological agent into a drinking water system. Under such circumstances a rapid and efficient concentration process is essential for the successful detection of biological agents. Recovery from large volumes of drinking water (10–100 L) was tested and optimized from several processes.

1. Optimization of model organisms recovery from 10 L of tap water. Recoveries were optimized using *Cryptosporidium parvum* oocysts, bacteriophages PP7 and MS2, a poliovirus (Sabin type 2 strain), *Escherichia coli*, and *B. globigii* spores:
   a. Optimization of the initial concentration step
   b. Optimization of downstream processes
   c. Determination of the storage feasibility of pretreated filters to reduce process time

2. Optimization of model organisms recovery from 100 L of tap water. Recoveries were optimized using *C. parvum* oocysts, and bacteriophages PP7 and MS2:
   a. Optimization of the initial concentration step
   b. Optimization of the downstream process
   c. Compatibility with infectivity and molecular detection methods

APPROACH

Hollow fiber ultrafiltration (UF) systems were used in the initial step of a multistep concentration process. The objective of the downstream concentration process was to further concentrate the sample after the initial UF process.

Two hollow fiber UF systems were tested for the recovery of viral, bacterial, and protozoan agents in tap water. The Microza hollow fiber UF system was chosen because it was extensively tested in ground- and surface water with several agents. These filters, however, were expensive. Initial experiments were performed with 10 L of dechlorinated tap water and then expanded to 100 L. Consideration for process conditions was given for simplicity, time, and the overall cost of the preconcentration procedure.

The concentration process included the initial pretreatment of the filters with a solution of 0.05% calf serum and 0.01% Tween 80. Calf serum was used to coat the surface of the membrane and reduce the level of adsorption of microbes to the surface of the membrane. Blocking of the filter surface by proteinaceous agents may also increase the efficiency of the elution process in order to resuspend agents associated with the surface of the membrane during filtration.

Experiments were included to test the feasibility of storing pretreated filters so that sample processing could occur immediately following sample collection. Filters were stored for 2 and 4 weeks in deionized water (after being blocked) or blocking solution with 0.01% thimerasol. The storage solution was tested for bacteria to determine whether bacterial contamination occurred during storage. A second filter was also tested (IMT hollow fiber ultrafilter), which is similar to the Microza filter but much less expensive.

Several downstream concentration methods (a secondary UF process, positively charged fibers and filters, spin columns) were tested to produce 10–20 mL of final concentrate. The objective was to produce a complete concentration process in less than 3 hours. Additional processes were needed to concentrate viral agents from the initial concentrate.

The use of a second hollow fiber UF method provided an additional UF-based approach for the concentration of viral pathogens. A much smaller filter module was used with 0.08 m^2 of surface area. Spin columns were then used in the final concentration process to reduce the final volume to 15–20 mL.

CONCLUSIONS

A UF-based preconcentration process for viral, bacterial, and protozoan agents in 10 L and 100 L of drinking water was developed and optimized. Recoveries were greater than 50% for bacteria and viruses, and greater than 25% for *C. parvum*. Specific conclusions include

1. Documentation of recoveries from 10 L and 100 L of drinking water using the Microza 50,000 molecular weight cutoff (MWCO) hollow fiber ultrafilter. The use of Tween 80 and sodium polyphosphate as amendments in the initial sample improved recovery efficiency as well as the use of 0.05*M* glycine at pH 7 as an eluent.
2. The storage of a blocked filter for up to 4 weeks was feasible with either sterile DI water or 0.01% thimerasol added to the blocking agent. No bacterial contamination was detected in the filter module during storage with either treatment. Storage of a blocked filter is a viable option.
3. The efficient concentration of viral, bacterial, and protozoan agents from 100 L of tap water to 15–20 mL was possible (>50% recoveries, except for *C. parvum* oocysts at ~25%. The process involved the use of two UF steps and the use of spin columns in the final concentration step. The entire process can be completed in less than 3 hours.
4. An alternate downstream concentration method is the use of positively charged fibers. The use of Dulbeccos minimum essential media with 10% fetal bovine or calf serum and sodium polyphosphate as eluents produced the most efficient recoveries. For viral agents, the use of adsorptive fibers and spin columns following the initial UF process offers an alternative concentration process.
5. The IMT hollow fiber ultrafilter offers a potentially lower-cost alternative to the Microza hollow fiber filter (IMT filters: $125 for 10 L and $275 for 100 L; Microza filters: $400 for 10 L and $1,500 for 100 L). Efficient recoveries were produced for 10-L volumes with the use of 0.1% sodium polyphosphate.

RECOMMENDATIONS

Concentration processes based on UF have been developed for 10 L and 100 L of drinking water and can reduce the volume to 15–20 mL within 3 hours. These research results and those of other recent studies have verified the utility of UF-based concentration processes for various water types.

The best results were obtained when 100 L of tap water was buffered with 10 L of phosphate-buffered saline and 0.01% Tween 80. The initial concentration process reduced the volume to 2 L. The secondary UF process condensed the concentrate to ~50 mL. The use of 30,000 MWCO spin columns further reduced the final volume to 15–20 mL.

The use of a less expensive hollow fiber ultrafilter showed promising results when tested in 10-L volumes. Larger modules that have the capacity to filter 100 L in 30 minutes should be tested.

Steps should be taken to further validate the process and to develop the hardware needed to produce the most user-friendly and safe operation in the field.

FUTURE RESEARCH

Additional study is being conducted to verify the compatibility of the concentrate with molecular methods as well as verify recoveries with low concentrations of model organisms. Use of 100 L of water with the IMT filters and a small number of organisms is also being conducted. The centrifugation step to pellet cellular organisms should be converted to a dead-end filtration approach to capture cellular organisms. This will simplify the process by eliminating the need for a centrifuge.

The major area of study needed to achieve the goal of a unified and rapid preconcentration process is the integration of the hardware necessary to produce either an in-line or field-deployable system. The system needs to be user-friendly and as automated as possible.

CHAPTER 1
INTRODUCTION AND BACKGROUND

The safety of drinking water sources and distribution systems is paramount to public health. Recent global events have highlighted the vulnerability of drinking water resources to accidental and intentional contamination with bio-threat agents. To ensure the safety of drinking water, source water and distribution systems must be monitored for the presence of waterborne pathogens.

The contamination of water systems with human pathogens is often difficult to assess, primarily because of the complexity of current methods to concentrate pathogens from large volumes of water. Many waterborne pathogens have low infectious doses, making them a public health risk even when highly dilute (Table 1.1). Comprehensive monitoring of drinking water for these pathogens requires a considerable concentration step that must effectively consolidate viruses, bacteria, and protozoa from large volumes of water.

Many approaches have been used to concentrate pathogens from water. Most of these methods have been developed for specific groups of organisms such as viral, bacterial, or protozoan agents (Table 1.2). Few studies have researched methods to simultaneously concentrate all three groups from large volumes of water.

Generally, there are three steps for detecting viruses in water: (1) collection of a representative water sample, (2) concentration of the virus in the sample, and (3) identification and quantification of the concentrated virus. The VIRus-ADsorption-ELution (Viradel) system, a standard method employed for concentration of viruses from water, is based on the adsorption and subsequent elution of viral particles from a charged filter membrane (APHA, AWWA, and WEF 1995; Goyal and Gerba 1982, Sobsey 1976). The method uses a negatively or positively charged 0.2–0.45-µm microporous membrane filter. For negatively charged filters, the water sample is conditioned to an acidic pH by using divalent (magnesium chloride, or $MgCl_2$) or trivalent (aluminum chloride, or $AlCl_3$) salts to alter the surface charge of the viral particles and enhance viral adsorption to the filter (Berg 1983). Water treatment is not necessary when using positively charged filters as they are able to efficiently bind viruses in the range of pH normally found in environmental samples (Berg 1983; APHA, AWWA, and WEF 1995). With negatively or positively charged filters, an alkaline eluent, such as glycine–sodium hydroxide (NaOH) or beef extract, is used to alter the viral charge and elute the virus into a much smaller volume (Berg 1983, Goyal and Gerba 1982).

Variation in water quality may affect the adsorption and elution of viruses from the filter membrane. Variation in recoveries can be due to differences in water chemistry (i.e., pH, salt concentration, presence of proteinaceous materials, humic compounds). These factors or a combination thereof have been shown to affect viral adsorption (Guttman-Bass and Catalano-Sherman 1985; Melnick et al. 1984; Oshima et al. 1995; Rose et al. 1984; Schwab, De Leon, and Sobsey 1995; Shields and Farrah 1983; Sobsey and Glass 1984; Sobsey and Hickey 1985). The presence of suspended solids in solution has been shown to interfere with the filtration process by clogging the filter or adsorbing and competing with viral particles for binding sites on the membrane. Virus particles may also adsorb to the suspended solids during the concentration process (Berg 1983; Dizer et al. 1993; APHA, AWWA, and WEF 1995; Sobsey and Glass 1984). The result may be the loss of the virus through the filter or binding of viral particles to suspended particulates, which makes viruses difficult to detect. The elution process has been shown to produce variable results

Table 1.1
Examples of viral, bacterial, and protozoan pathogens transmitted though water

Group	Pathogen	Size	Symptoms and diseases
Viruses	Enteroviruses: polio-, echo-, coxsackie-	30 nm	Diarrhea, rash, fever, respiratory disease, paralysis, meningitis, myocarditis
	Hepatitis A virus	27 nm	Diarrhea, infectious hepatitis
	Norovirus	30 nm	Diarrhea, gastroenteritis
	Adenovirus	90 nm	Diarrhea, eye infection, respiratory illness
	Rotavirus	60 nm	Diarrhea, gastroenteritis
Bacteria	*Escherichia coli*: O157:H7 and others	1 × 3 μm	Diarrhea, hemolytic uremia syndrome
	Salmonella	~1 μm	Diarrhea, typhoid
	Shigella	~1 μm	Diarrhea
	Legionella	~1 μm	Respiratory infection, pneumonia
Protozoa	*Cryptosporidium parvum*	4 μm	Acute or chronic diarrhea
	Giardia lamblia	9–12 μm	Chronic diarrhea
	Entamoeba histolytica	15–60 μm	Amoebic dysentery
	Cyclospora spp.	8–10 μm	Diarrhea

Table 1.2
Current standard and developing methods used for concentration and isolation of pathogens from water

Method	Agent	Type of filtration	Initial water volume (L)	Pore size (μm)	Downstream processes
Standard	Virus	Microfiltration	10–1,000	0.2	Polyethylene glycol; organic flocculation
	Protozoa	Microfiltration	10	1.0	Centrifugation
	Bacteria	Microfiltration	0.1	0.4	None
Current study	Virus, protozoa, bacteria	Ultrafiltration	10–100	~0.02	Centrifugation, filtration, additional viral processes

based on the virus type and water quality (Berg 1983; Dizer et al. 1993; APHA, AWWA, and WEF 1995; Sobsey and Glass 1984). Furthermore, the Viradel method has not been optimized for the concentration of bacteria or protozoa.

Following the elution process, additional procedures such as a secondary filtration process, polyethylene glycol hydroextraction, dialysis, organic flocculation, and aluminum hydroxide flocculation can be used to further concentrate the sample to <10 mL (Berg 1983; APHA, AWWA, and WEF 1995; Goyal and Gerba 1982).

Processes may also need to incorporate the removal of toxic compounds and microbial contaminants prior to cell culture. These purification procedures are necessary because compounds found in water, such as humic and fulvic acid, are concentrated along with viral particles, and can cause toxic effects in cell culture or can inhibit molecular methods (Wilson 1997). Procedures to remove these inhibitors include immunomagnatic separation (IMS), Sephadex and Chelex spin columns (Sigma, St. Louis, Mo.), and protein precipitation (Graff, Ticehurst, and Flehmig 1993; Jothikumar, Cliver, and Mariam 1998; Lantz et al. 1997; Schwab, De Leon, and Sobsey 1995).

Standard methods for the detection of waterborne protozoa (*Cryptosporidium parvum* or *Giardia lamblia*) require the use of a 1-µm-pore size filter to concentrate oocysts from the sample as well as a subsequent elution step to remove bound protozoa from the filter. IMS removes particulates from the sample, and indirect immunofluorescent antibody (IFA) techniques are used to visualize the oocysts by fluorescent microscopy for quantification (USEPA 2001b).

For bacterial isolation and enumeration, the most commonly used method is to filter 100 mL of water through a 47-mm diameter, 0.45-µm pore size, cellulose ester membrane filter. This is the standard method for detection of total and fecal coliforms in water (USEPA 2002a, b, c).

Filtration with hollow fiber ultrafilters offers an alternative concentration method that can simultaneously concentrate all types of pathogenic organisms (Belfort, Rotem, and Katzenelson 1975; Bicknell, Dziewulski, and Sturman 1985; Green and Tylla 1998; Juliano and Sobsey 1998; Kfir et al. 1995; Kuhn and Oshima 2002; Morales-Morales et al. 2003; Olszewski, Winona, and Oshima 2001, 2005; Payment et al. 1989; Simmons et al. 2001). Hollow fiber ultrafilters are cartridges packed with many individual strands of hollow membrane. The hollow tubes of membrane material are designed with a porous outer layer containing a semipermeable coating on the inside, arranged in bundles of tubes. Ultrafilters use membranes with pore sizes ranging from 0.02 to 0.1 µm and are commonly composed of noncellulosic synthetics (such as polyvinylidene fluoride, polyacrylonitrile [PAN] or polysulfone) that are resistant to heat and chemical attack. Many are designed for low protein-binding and are reusable in many cases (Brock 1983).

In crossflow mode, as the water sample passes through the interior of the strands, pressure is applied to push water across the membrane (through the pores) of the strands while particulates larger than the MWCO value remain suspended in the concentrate. Pressure is maintained at a level that allows only a portion of the water to filter through, and the remaining water is allowed to flow back to the reservoir. The "clean" water that passes across the hollow fiber membrane is termed the "permeate" and should be mostly free of organisms. The recirculated sample is termed the "retentate" and should contain a large percentage of the agents that were in the original suspension. The sample is continually recirculated through the ultrafilter cartridge until only the holdup volume in the system remains.

Standard microfiltration methods bind microorganisms on a filter and rely on subsequent elution steps to recover target organisms. Often, this multistep binding and elution is detrimental to the organisms as well as inefficient in recovery. Any unbound agents (particularly viruses) will pass through the filter pores and be lost during the process. All of the bound organisms may not be eluted off of the filter membrane, adding to the potential loss of recovery efficiency.

Ultrafiltration (UF) systems use size exclusion to push water across the membrane while prohibiting the passage of organisms. This process, together with differences in membrane construction, may allow organisms to remain suspended in the retentate or may allow for more efficient elution of bound microbes from the membrane surface. Crossflow mode of hollow fiber UF systems also reduces fouling (accumulation of a gel layer) of the filter membrane by providing constant flow through and across the membrane. Dead-end filtration systems are particularly susceptible to membrane fouling because of the accumulation of suspended organic material and particulates.

Ultrafilters are designed to be reusable following a sanitation process that can include NaOH, hydrochloric acid (HCl), bleach, and sodium dodecyl sulfate (SDS), or multiple washes with these agents. These cleaning agents remove proteins, minerals, and polysaccharides from the membrane to prevent fouling while disinfecting the filter to remove and inactivate microbial agents. Because pathogens transmitted through water commonly have low infectious doses, these agents represent a substantial public health risk even when highly dilute. The use of 100-mL samples in standard methods for the detection of bacteria from water is inadequate for the detection of dilute pathogens in water. Effective monitoring of drinking water requires the assessment of volumes on a scale of 100–1,000 L for the presence of target organisms. Furthermore, these large volumes must be reduced to a very small concentrate to maximize the sensitivity of the final assays.

Further complicating the evaluation of drinking water safety is the issue of the time necessary to process the sample. In the event of contamination of a drinking water supply, there must be a method in place that allows rapid processing and identification of the specific threat, thereby enabling actions to limit public exposure to contaminated drinking water.

ALTERNATIVE APROACHES TO CONCENTRATE MULTIPLE PATHOGENS FROM WATER

Many approaches have been used to concentrate pathogens from water. Most of these methods have been developed for specific groups of organisms such as viral, bacterial, or protozoan agents. Fewer studies have investigated ways to simultaneously concentrate all three groups from large volumes of water.

UF (tangential flow or hollow fiber) is a technique that can be used for the simultaneous concentration of diverse waterborne microbes in a single step (Kfir et al. 1995, Hill et al. 2005, Morales-Morales et al. 2003, Oshima 2001, Juliano and Sobsey 1998).

Variation in the efficiency of recovery among the different types of pathogens has been reported but processing of large volumes often takes too long (Juliano and Sobsey 1998, Kfir et al. 1995, Morales-Morales et al. 2003). Hollow fiber systems filtered 50 L of tap water in approximately 3.5 hours and filtered 10 L in 1.5 hours (Juliano and Sobsey 1998; Simmons et al. 2001).

Recent research by Morales-Morales et al. (2003) reported >50% recoveries for bacteriophages *Escherichia coli* and *C. parvum* oocysts seeded into 10-L surface water samples. Their method, however, was not developed as a rapid method and used an overnight pretreatment of the filter with calf serum to reduce adsorption of microbes to the filter surface during filtration. For bio-threat response applications (e.g., rapid response and field-based filtration), pretreatment with calf serum may not be as desirable because of the added time involved in the overall process.

Properties of microbial particles, such as size, shape, composition of outermost layer, and stability, have been shown to influence concentration efficiency (Juliano and Sobsey 1998, Simmons et al. 1999, Sobsey 1976, Venczel et al. 1997). The recovery of organisms is also affected by water quality parameters such as turbidity, pH, levels of salts, and organics (Gerba et al. 1978, Guttman-Bass and Catalano-Sherman 1985, Kuhn and Oshima 2002, Sobsey 1976). UF (Divizia, Santi, and Pana 1989; Juliano and Sobsey 1998; Kfir et al. 1995; Morales-Morales et al. 2003) and microfiltration (Kfir et al. 1995, Payment et al. 1989) have been used to concentrate multiple organisms from water.

Considerable work has been done in the development of methods to concentrate microbes with the Microza hollow fiber ultrafilter (Pall Corporation, Port Washington, N.Y.). The ultrafilter has been used to concentrate organisms from surface, ground-, tap, and raw waters as well as treated sewage (Kuhn and Oshima 2002; Olszewski, Winona, and Oshima 2001, 2005; Morales-Morales et al. 2003). Many bacterial, viral, bacteriophage, and protozoan agents have been used. Initially, experimentation was prepared in small volumes (2 L) to optimize recovery and was then scaled up to field-size volumes (10–100 L). Small volumes could be tested to model recoveries in field-scale systems, and recoveries from up to 90 L are similar to the smaller volumes even when turbid samples are used (Kuhn and Oshima 2002; Olszewski, Winona, and Oshima 2001, 2005).

The UF methodology was applied to the concentration of bacterial, viral, and parasitic organisms (in 10 L and 100 L) in the Rio Grande, which at times can be a highly turbid river with nephelometric turbidity units averaging between 50 and 60 and reaching as high as 650. This study concluded with the following observations:

1. Recovery by size selection of viruses, parasites, and bacteria can be performed using the same initial concentration method.
2. Recoveries appear to be consistent between widely varying organisms and water qualities.
3. Larger volumes of water may be sampled compared to existing methods (increase in detection sensitivity).
4. The filters can be reusable and thus are more economical, and there is a consolidation of methods that can reduce cost.
5. The overall concentration scheme is simpler compared to existing methods.
6. UF may also offer advantages over the standard method in the detection of organisms when molecular methods are used (i.e., the polymerase chain reaction).

Adsorptive Microfilters

Nano alumina fibers attract and retain viruses and other macromolecules by electrostatic forces. These fibers have been incorporated into fibrous ("depth") filters and have approximately 7–8 logs of virus retention even at flow rates several orders of magnitude greater than could be obtained using traditional microporous membranes. Boehmite (AlOOH) is the main component of the fibers. The large number of hydroxyl groups available on the nanofibers generates a positive charge in water solution such that it will attract and retain negatively charged particles including bacteria, viruses, protozoa, organic and inorganic colloids, and negatively charged macromolecules. This type of filter is fundamentally different than the ultrafilters. Viruses are concentrated by adsorption to the material whereas cellular pathogens are retained by size selection. The fibers can also be added on to a surface of a supporting material at different depths or added to a solution as a powder.

Filters are classified as fibrous depth, surface, or membrane filters. One type of filter, the NanoCeram filter (Argonide Corp., Sanford, Fla.), is based on fibers and therefore would be classified as a depth filter. Bubble pressure tests on the NanoCeram filters show that although the apparent pore size is in the range of 0.8 µm, they are capable of high retention for much smaller particles and, in fact, retain micromolecules with dimensions of only a few nanometers.

Disposable Hollow Fiber Ultrafilter

The disposable hollow fiber ultrafilter is a polyethersulfone UF membrane. One of the major advantages of the ultrafilter is the projected cost as a single-use, disposable system. The lower cost makes it feasible to use once and then discard, eliminating the need to sanitize and disinfect the filter after use. The concentration process would then be greatly simplified and more rapid. However, this filter could also be used as a reusable filter if cost at approximately $240 per module became an issue.

Preliminary studies consisted of bacteriophage PP7 challenges to determine the best MWCO for size selection based on virus retention. Based on these experiments, the MWCO for retaining phage PP7 and maintaining a desirable flow rate was 40,000–50,000 Da.

Various virus recovery experiments were then initiated to begin to optimize recovery. All experiments were conducted with phages T1 and PP7. Several elution and backflush experiments were performed, and the recoveries were encouraging.

SELECTION CRITERIA FOR RAPID EXTRACTION METHODS

The objective of this project was to develop methods for the rapid and simultaneous extraction of a wide variety of pathogens from drinking water when confronted with the potential intentional introduction of a biological agent. Under these circumstances, a rapid and efficient concentration process is essential for the successful detection of the biological agent. The following major criteria were used in the selection of approaches for the concentration process:

1. Recovery of >50% of the spiked model organisms
2. Completion of the process within 3 hours
3. Simplicity of the process
4. Ability to handle wide ranges of water qualities
5. Cost

Estimated cost of each filter or media (single-use costs):

	10 L	100 L
Microza hollow fiber	$440	$1,500
Adsorptive media	<$10	$10–$20
Innovative Membrane Technologies (IMT; Mast, Netherlands) hollow fiber	$125	<$275

OBJECTIVES

Recovery from large volumes of drinking water (10–100 L) was tested and optimized from several processes:

1. The optimization of recovery of model organisms from 10 L of tap water. Recoveries were optimized using *C. parvum* oocysts, bacteriophages PP7 and MS2, poliovirus (Sabin type 2 strain), *E. coli*, and *B. globigii* spores:
   A. Optimization of the initial concentration step
   B. Optimization of downstream processes
   C. Determination of storage feasibility of pretreated filters to reduce process time

2. The optimization of model organisms recovery from 100 L of tap water. Recoveries were optimized using *C. parvum* oocysts, and bacteriophages PP7 and MS2:
   A. Optimization of the initial concentration step
   B. Optimization of the downstream process
   C. Compatibility with infectivity and molecular detection methods

CHAPTER 2
MATERIALS AND METHODS

The organization of this chapter primarily follows the sequential order in which the project was conducted. Initial work focused on the use of the Microza hollow fiber ultrafilter and reproduction of previous results from surface and groundwater sources for drinking water in 10-L samples. The concentration process then proceeded to using 100-L drinking water samples with the primary objective of optimizing recoveries from the initial UF process. The study also focused on the development of storage procedures for blocked filters. Additional first-step concentration procedures, such as a secondary ultrafilter (from IMT) and the use of adsorptive microfiber filters, were also investigated.

The final stages of the project focused on additional concentration procedures for the concentrate of the initial UF process to further address concentration procedures for viral agents. Methods included a secondary UF process, and the use of adsorptive fibers and spin columns. When a complete process was identified, the final stage was to test that process.

10-L HOLLOW FIBER ULTRAFILTRATION

Filter Preparation

The hollow fiber UF system consists of a filter, a tubing system, two reservoirs, and a peristaltic pump (Kuhn and Oshima 2002; see Figure 2.1). PAN 50,000 Da MWCO Microza ultrafilters with a surface area of 0.2 m^2 were used on 10-L volumes of tap water. The membrane was sanitized (new and reused) by recirculating a 500-mL solution of 200 mg of sodium hypochlorite and 0.5*M* NaOH for 30 minutes. The sanitation solution was rinsed from the module by filtering reverse osmosis (RO) water through the filter for 30 minutes (~100 L). The free chlorine level was measured using N,N-diethyl-p-phenylenediamine (DPD; DR/2010 method ID number 80; Hach Chemical Company, Loveland, Colo.). The module was flushed with RO water until the residual free sodium hypochlorite was below 0.02 mg/L. The filter was treated overnight with 10% SDS and incubated at 37°C to prevent carryover of *Cryptosporidium* and *Giardia* by the induction of excystation and inactivated sporozoites (Kuhn and Oshima 2002). (Detailed protocols for this procedure are provided in appendixes A and B.)

Schemes for Blocking the Filter

When sanitation was completed, the module was blocked (pretreatment of regenerated ultrafilters with proteinceaous agents to reduce the binding of microbes to the filter surface) with 500 mL of 5% calf serum. The blocking solution was circulated in the module in crossflow mode only (no permeate flow) for 1 hour on the day before the experiment. The module was then placed on a shaker at 4°C for 16–18 hours. The following day, the blocking solution was drained and the module rinsed with 4 L of RO water in the crossflow mode (single pass).

In separate experiments, the effect of storage on blocked filters for up to 1 month was discovered. Two types of storage conditions were tested. The first set of experiments determined the effect of storage of blocked filters in sterile deionized (DI) water without thimerasol, and the second set determined the storage of filters in 0.05% calf serum with 0.01% thimerasol. After

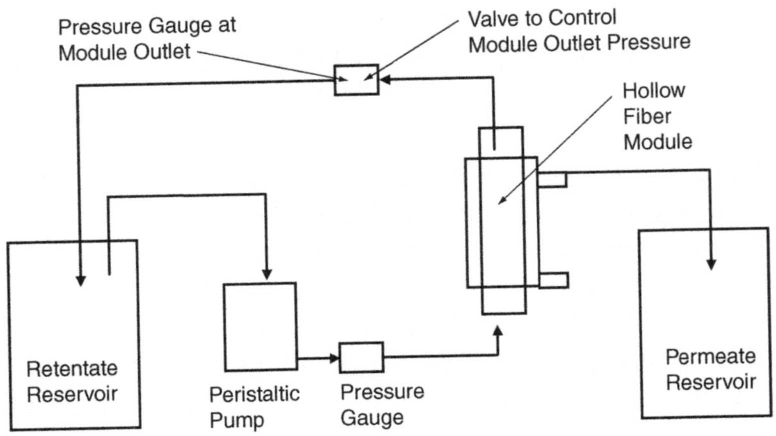

Figure 2.1 Diagram of filtration scheme for the hollow fiber filtration systems

storage, in both cases, the filters were drained according to the standard procedure. While the filter was stored, a sample of the solution was tested for bacteria twice per week to determine possible bacterial contamination during storage (appendix C). A minimum of three replicated recovery experiments per microbe were conducted for blocked filters that were stored in DI water and for filters that were blocked and stored with 0.1% thimerasol.

Filtration of Tap Water

Ten liters of tap water was collected and 50 mg/L of sodium thiosulfate was added. The free sodium hypochlorite was measured as previously described. The water was used only if the free chlorine level was less than 0.02 mg/L. Tap water (from a groundwater source) collected from Foster Hall at New Mexico State University was used as the water source for all experiments. The test organisms (bacteriophages PP7 and MS2, poliovirus, *C. parvum* oocysts, *B. globigii,* and *E. coli* K-12) were then added in different combinations depending on the experiment (approximately 100–1,000 organisms/mL). Tween 80 (0.1% final concentration; Sigma) and 10× phosphate-buffered saline (PBS) were also added to the 10-L sample and mixed. A 10-mL sample of the suspension was then collected to determine the starting concentration, and the filtration process was started (400 mL/min permeate flow at <15 psi). The filtration was terminated when the retentate volume was approximately 200 mL, and the duration of filtration was recorded.

Twenty-five milliliters of 0.5*M* glycine was added to the retentate sample, and the retentate volume was adjusted to 250 mL with RO water. Next, the retentate solution was recirculated (crossflow mode) through the filter for 30 minutes. The sample was then centrifuged at 1,240× gravity (*g*) for 20 minutes. The pellet was sampled for cellular organisms and the supernatant tested with viral assays.

Regeneration of the Filter

The sanitation process was repeated after filtration and elution were completed. If necessary (every fourth use), an acid wash was also performed with 1% HCl in 500 mL and recirculated (crossflow mode) for 1 hour. The acid from the module was rinsed with RO water for 30 minutes or until ~100 L was passed through the filter. Next, a clean water flux sample was collected to

Table 2.1
Description of model organisms used for tap water filtration experiments

Organism	Type	Purpose	Size	Shape
Phage PP7	RNA phage	Surrogate for enterovirus	25 nm	Icosahedral, nonenveloped
Phage MS2	RNA phage	Surrogate for enterovirus	25 nm	Icosahedral, nonenveloped
Poliovirus Sabin type 2	RNA virus	Model enterovirus	27 nm	Spherical, nonenveloped
E. coli K-12	Bacteria cells	Vegetative bacterial cells	1 × 3 μm	Rod-shaped
B. globigii spores	Bacteria spores	Surrogate for *Bacillus anthracis* spores (anthrax)	1 μm	Spore
C. parvum	Protozoa	Model protozoan parasite	4 μm	Spherical

determine whether the flow rate met the pre-use flux. If the flow was adequate, the module was then stored with 50 mL of 10% SDS solution in a 37°C incubator overnight. The next morning, the SDS was removed by rinsing for 20 minutes with RO water in the crossflow mode.

Microorganisms

Viral, bacterial, and protozoan agents were used in this study (Table 2.1; see details in appendixes B and D). *C. parvum* oocysts were purchased from Waterborne Inc. (New Orleans, La.). Stocks were diluted in 1× PBS, labeled with IFA (Waterborne) and enumerated with an epifluorescent microscope prior to spiking of water samples. Oocysts were isolated from the resuspended concentrates using IMS (Aureon Crypto kit from ImmTech, New Windsor, Md.). Oocysts were then labeled with an IFA and visualized with an epifluorescent microscope.

E. coli K-12 was obtained from the New Mexico State University Biology Department Culture Collection. *E. coli* K-12 was grown in nutrient broth at 37°C on a shaker functioning at 200 rpm for 17 hours (optical density reading of 0.8–0.9). This bacteria was assayed by a drop-and-spread technique with 0.1 mL dropped onto MacConkey agar plates.

B. globigii spores were supplied by the New Mexico State University Biology Department Culture Collection. *B. globigii* spores were enumerated using 0.1 mL volumes for the drop-and-spread technique on tryptic soy agar plates. HeLa cells were used for the growth and enumeration of poliovirus type 2 (Sabin strain) (Oshima et al. 1995).

Phage PP7 (American Type Culture Collection [ATCC] 15692-B2) and phage MS2 (ATCC 15597-B1) were cultivated in host bacteria *Pseudomonas aeruginosa* (ATCC 15692) and *E. coli* F_{AMP}, (ATCC 15597), respectively. Poliovirus and phage PP7 were enumerated using the plaque assay techniques described in other publications (Oshima et al. 1995). Phage MS2 was cultivated and enumerated according to U.S. Environmental Protection Agency (USEPA) method 1602 (USEPA 2001a). Volumes of 1-mL, 0.5-mL, and 0.1-mL were used for phage PP7, phage MS2, and poliovirus assays, respectively. All assays were plated in duplicate.

Phage PP7 was used as a control to monitor recovery and filter integrity. For this reason, the phage was added to all experiments in this study.

Viruses and bacteria were added to 100-L drinking water samples to attain an initial concentration of approximately 1×10^6 cfu or pfu per liter. The target concentration for *C. parvum* was approximately 100 oocysts/L of drinking water.

STORAGE OF BLOCKED FILTERS

The effect of storage of blocked filters for 2 and 4 weeks on microbial recovery was also tested and compared to filters that were not stored (appendix C). Stored filters were evaluated for bacterial contamination.

Recoveries of selected organisms were determined when blocked filters were stored for 2 and 4 weeks at 4°C, and three replicated experiments were conducted for each organism. Filters were stored in two ways. In the first method, filters were blocked by the standard method. The module was then drained, and 200 mL of sterile DI water was added to the module by removing one of the end caps, pipetting the DI water, and replacing the end cap. Periodically, a sample was collected from a tube that extends from another tube at the end of the module. The tube was unsealed, and a 5 mL sample was collected. The second approach stored filters in the blocking solution (5% calf serum with 0.01% thimerasol). Thimerasol was added to reduce the potential for bacterial regrowth during storage. In both methods, sterility of the filter modules was checked for the presence of heterotrophic bacteria. At the end of the storage period, the filters were rinsed, as described in appendix A for unstored filters.

To determine the effectiveness of the preservative, the filter modules were analyzed for bacterial contamination twice per week.

100-L TAP WATER CONCENTRATION PROCESS

After methods were verified for 10-L samples, the next set of experiments evaluated the methods for 100-L samples. A larger-capacity filter (1 m^2 of surface area) was used, as well as a larger peristaltic pump and tubing system. After the completion of optimization processes, additional concentration methods were studied to reduce concentrate for viral analysis to 10–15 mL.

Preparation of Filters

The procedure for preparing 100-L filters is very similar to the 10-L method. The main differences were the volumes of the various reagents and wash steps, and the flow rate of the filtration process. Prior to filtration, the ultrafilters were sanitized with a solution of 0.5*M* NaOH with 5% sodium hypochlorite (v/v). The sanitant was circulated through the filter cartridge for 60 minutes in the crossflow mode. The filter was then rinsed with 40 L of RO water (permeate port open), dechlorinated with a 0.1% sodium thiosulfate solution in crossflow mode for 60 minutes, and rinsed again with 40 L of RO water (permeate port open). After the rinse step, to ensure that there was no residual sanitant in the filter, the permeate was tested to verify that the pH measured 7–8 (matching the pH of the RO water) and the chlorine was below 0.02 ppm. Free chlorine was measured with the DPD method (Method ID number 80) on a Hach DR/2000 spectrophotometer. The large ultrafilter was blocked by circulating 5% calf serum solution (v/v) in crossflow mode for 60 minutes. The filter cartridge was then filled with 5% calf serum and incubated overnight at 4°C on an orbital shaker. Prior to filtration, this module was rinsed with 20 L

RO water. The pH and chlorine level of the permeate was checked to ensure complete elimination of sanitant from the filter cartridge.

After sanitization, the module was filled with 0.5*M* NaOH, 5% sodium hypochlorite (v/v) solution and stored for up to 24 hours prior to use. The cartridge was rinsed immediately before use with ~10 L RO water (permeate port open) to eliminate remaining chlorine or base from the sanitizing solution.

Preparation of Spiked Drinking Water Sample

One hundred liters of drinking water was collected from the tap source (Foster Hall, New Mexico State University) and dechlorinated overnight with 10 g of sodium thiosulfate. Prior to introduction of biological agents, 10 L of 10× PBS was added to the water. The sample was then amended with 0.01% Tween 80. For some experiments, 0.01% NaPP (sodium polyphosphate) was also added. Calf serum was used as an amendment for some experiments.

Ultrafiltration Process

One-hundred-liter water samples were concentrated using a 50,000 Da MWCO, large-scale hollow fiber ultrafilter with 1.0 m^2 surface area of PAN membranes (AHP-2013 Microza) with a Masterflex high-performance pump (77600-02; Cole-Parmer, Vernon Hills, Ill.) and Masterflex Pharmed tubing (number 6485-89, Cole-Parmer; see Figures 2.2 through 2.4).

The water sample was circulated through the large ultrafilter for 10 minutes in crossflow mode (closed permeate port) to further mix the sample. Next, the pump was stopped, and all water was drained back into the retentate reservoir before an input sample was collected.

The permeate port was then opened and the pump was set at speed 7. Backpressure was adjusted to ~12 psi to reach a target initial permeate flow rate of 5,000 mL/min. The resulting permeate was collected in a separate reservoir while the retentate was reduced to ≤1000 mL. The permeate port was then closed. A 0.5*M* glycine pH 7 solution was added to the concentrate to attain a glycine concentration of 0.05*M*. The concentrate with glycine was next recirculated through the ultrafilter in crossflow mode for 30 minutes, and then centrifuged at 1,240× *g* for 20 minutes (Jouan CR3I; Jouan Inc., Winchester, Va.). The resulting supernatant was either subjected to further concentration processes or stopped at this point because the experiments were designed to test only the initial UF process.

The supernatant from the initial UF was partially removed, leaving a pellet volume of 5–10 mL. A vortex was created for this volume to resuspend the pellet and divided for cellular assays. A 0.5-mL volume was used for bacterial assays while the rest of the pellet volume, usually around 5–10 mL, was used for protozoan detection described later.

The 100-L experiments were conducted with the standard amendments (Tween 80) and also with Tween 80 and NaPP. Experiments with each organism were repeated at least three times.

Two-Step Ultrafiltration Process

Concentration Process

One hundred liters of drinking water was concentrated to ~20 mL using two UF steps, followed by spin columns. First, the 100-L tap water sample (100 L tap water and 10 L 10× PBS)

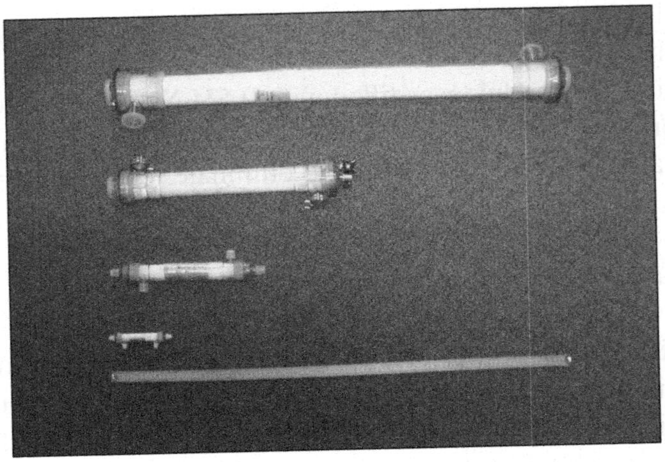

Figure 2.2 Microza ultrafilter modules. The largest module (top) is capable of filtering 1,000 L of water; next (second from top) is the 100-L filter; then the 10 L; and (bottom) the filter used for the secondary UF step.

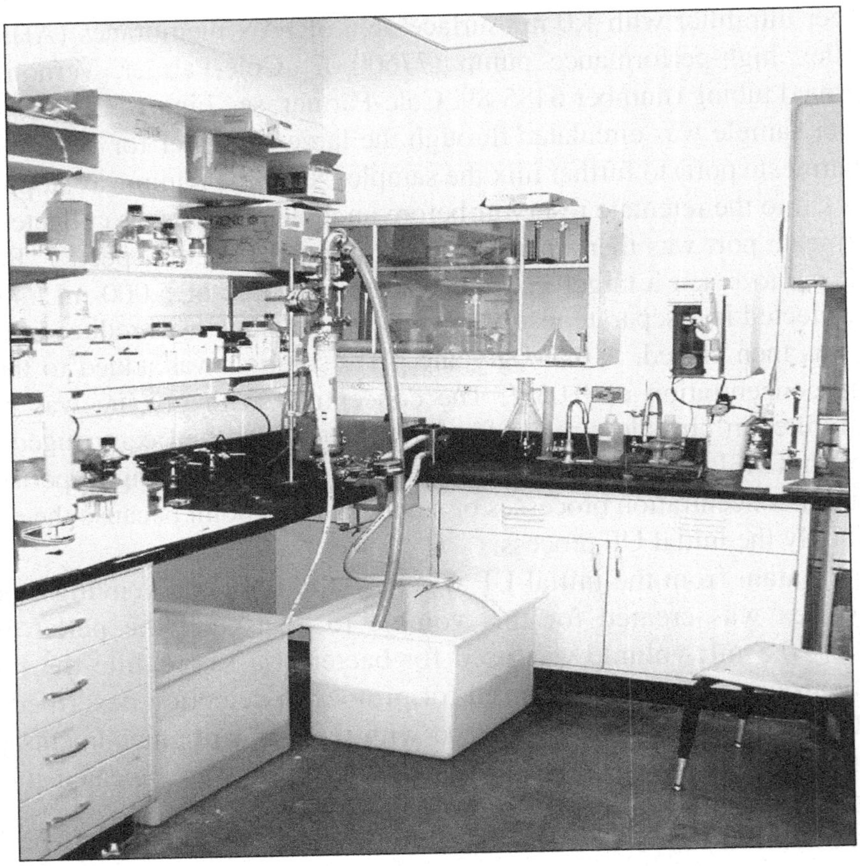

Figure 2.3 100-L large-scale UF system

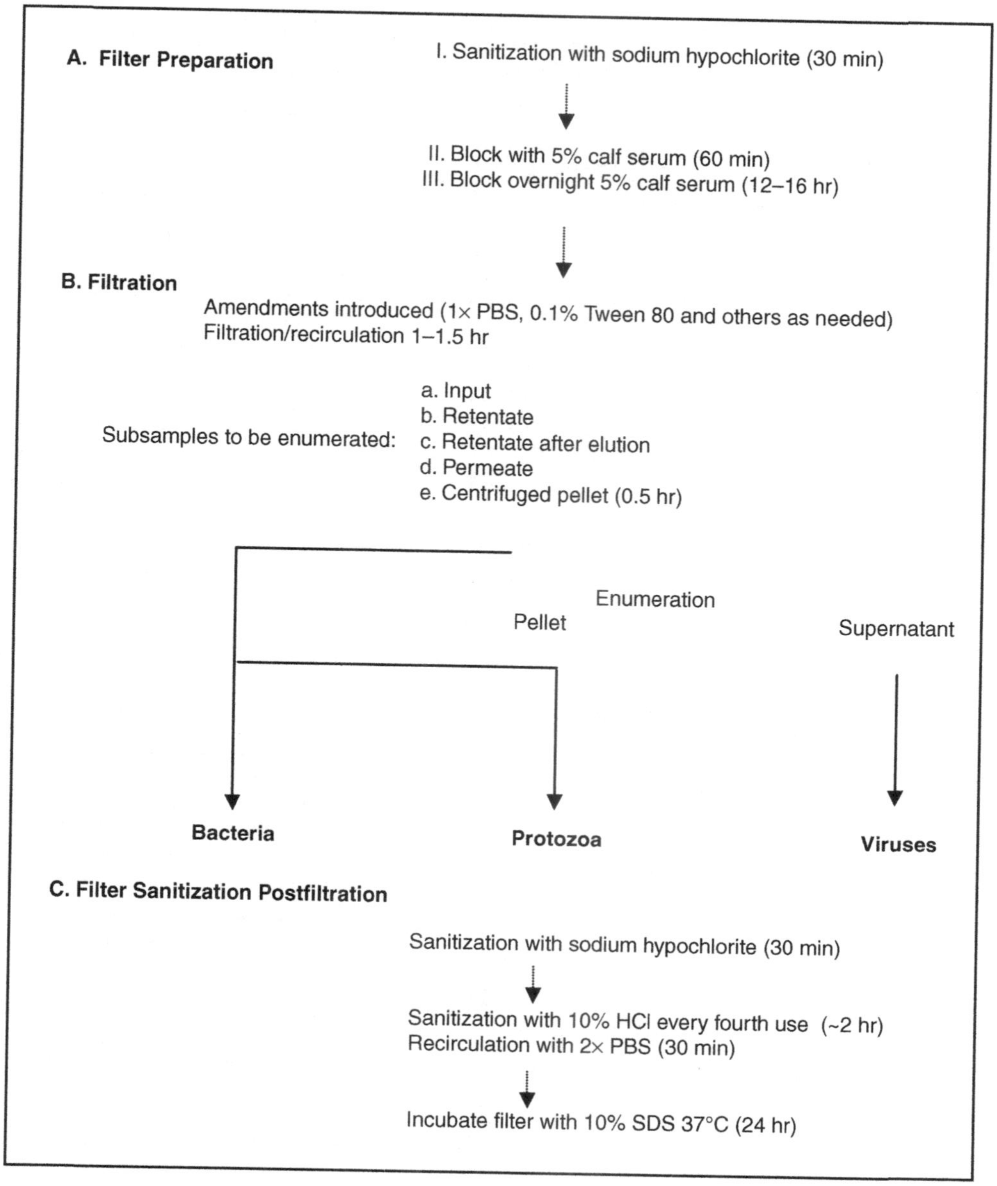

Figure 2.4 Standard method for the reusable hollow fiber ultrafilter (100 L) for concentrations of viral, bacterial, and protozoal pathogens from water

was concentrated to ~1 L using a large hollow fiber ultrafilter. The resulting concentrate was then centrifuged. The pellet was directly used to isolate cellular organisms. The supernatant was further reduced to a volume of ~50 mL using a smaller hollow fiber ultrafilter. Spin columns were used to further concentrate the resulting retentate to ≤20 mL. This final concentrate was used for the detection of viral agents (Figure 2.5).

Preparation of Filters

The filters were prepared in the same manner as previously described.

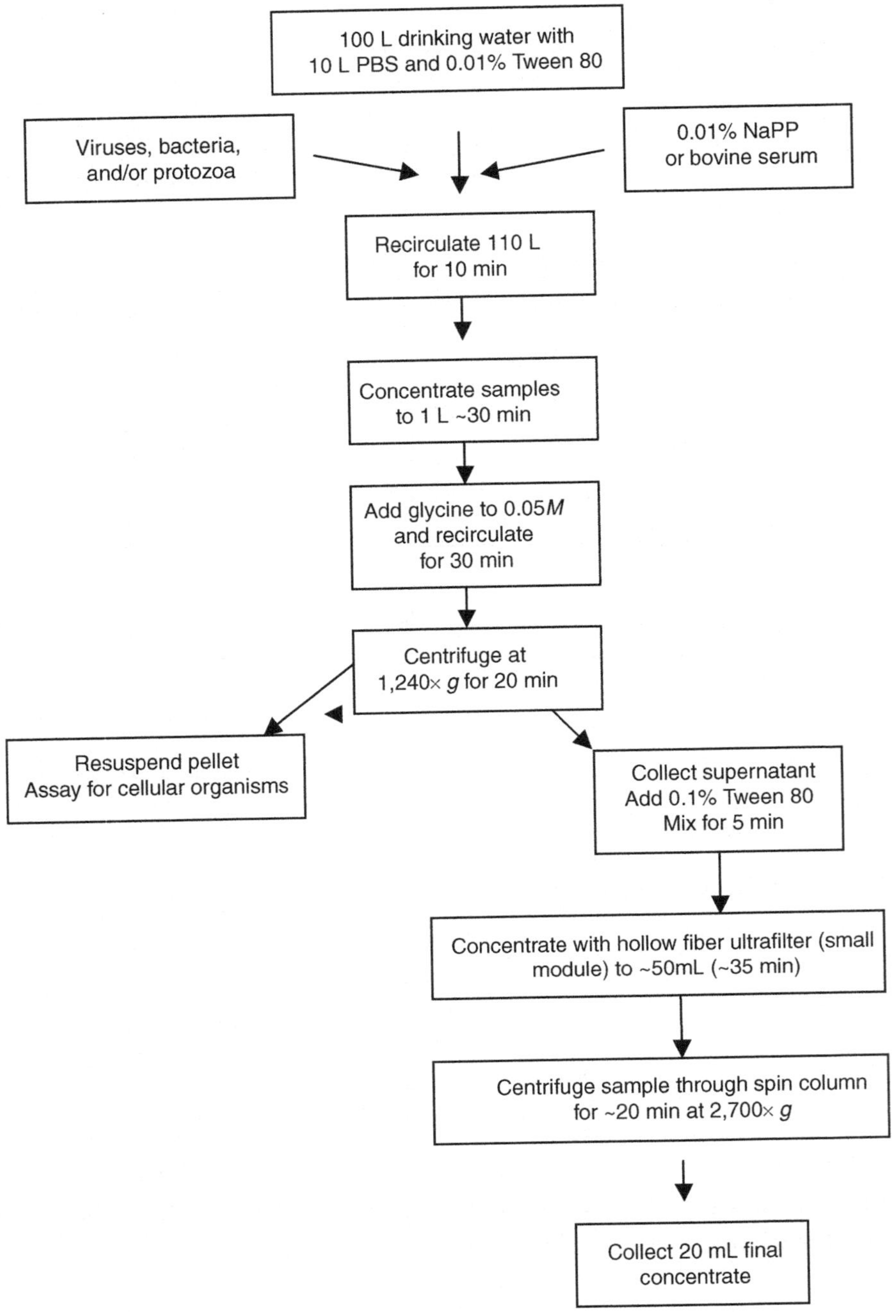

Figure 2.5 Flow chart of the two-step UF process to concentrate a 100-L sample

Preparation of Spiked Drinking Water Sample

The spiked 100-L drinking water sample was prepared in the same manner described previously. Appropriate amendments were added (Tween 80 and NaPP, depending on the experiment). The resulting 110-L sample was mixed, and appropriate biological agents were added.

Primary Ultrafiltration

One-hundred-liter water samples were concentrated using a 50,000 Da MWCO ultrafilter. The resulting permeate was collected in a separate reservoir while the retentate was reduced to ≤1,000 mL. The permeate port was then closed. A 0.5*M* glycine pH 7 solution was added to the concentrate to attain a glycine concentration of 0.05*M* in the concentrate. The concentrate with glycine was then recirculated through the ultrafilter in crossflow mode for 30 minutes, and centrifuged at 1,240× *g* for 20 minutes (Jouan CR3I). The resulting supernatant was subjected to a secondary UF process whereas the pellet was used for detection of cellular organisms.

The supernatant from the initial UF was partially removed, leaving a pellet volume of 5–10 mL. A vortex was created for this volume to resuspend the pellet, and then the volume was divided for cellular assays. 0.5 mL was used for bacterial assays whereas the rest of the pellet volume, usually around 5–10 mL, was used for protozoan detection.

Secondary Ultrafiltration

The supernatant from the centrifugation process was further concentrated using a 50,000 MWCO PAN hollow fiber ultrafilter with a surface area of 0.017 m^2 (AHP-0013, Microza) installed into an integrated pump system (PS 2400, Pall Corp.). The sample was amended with 0.1% Tween 80 (final concentration) prior to the secondary UF step. The sample was concentrated to ~50 mL using a transmembrane pressure of 80 kPa as described in other publications (Kuhn and Oshima 2002, Winona et al. 2001). Because the sample was under prime elution conditions (presence of Tween 80 and glycine in the concentrate), no recirculation of the concentrate was necessary.

Spin Columns

The resulting retentate was further concentrated with spin columns (Jumbosep, Pall Corp.), centrifugal devices with a modified polyethersulfone filter disc (30,000 Da MWCO; 15.2 cm^2 surface area) to ≤20 mL. The sample was centrifuged at 2,650× *g* (Jouan CR3I) for 10-minute increments until the final concentrate volume was less than or equal to 20 mL. An elution step was not needed because the concentrate already contained Tween 80 and glycine. The final concentrate was then used for viral analysis.

Samples Collected for Microbial Assessment

Samples were collected from various stages within the concentration process. For each experiment, an input sample (10 mL) was collected after the 10-minute recirculation step prior to large-scale UF. After the initial UF and recirculation with glycine, a primary retentate sample (10 mL) was collected, representing the first UF step recoveries. Pellet (total volume used for detection of bacteria and protozoa) and supernatant (10 mL) samples were collected after the centrifugation step. After the secondary UF, a 2-mL sample was collected representing the secondary retentate. After spin column centrifugation, the final concentrate was tested for viral analysis. To confirm the integrity of the ultrafilters, permeate samples were collected after both the primary and secondary UF steps and tested for all organisms used except *C. parvum*. All calculations accounted for loss of volume due to sampling.

Microorganisms

The same organisms and assays were used except for formalin-fixed *Cryptosporidium* oocysts as a model cyst-form protozoa. Formalin-fixed *C. parvum* oocysts were purchased from Waterborne Inc.

Calculations

Cumulative percent recovery was determined by dividing the total number of organisms recovered from the specific step by the total number of organisms in the initial drinking water sample (Equation 2.1).

$$\text{cumulative percent recovery} = \frac{\text{total number of organisms in sample}}{\text{total number of organisms in intital water sample}} \times 100 \qquad (2.1)$$

Stepwise recoveries were determined by dividing the number of organisms recovered from the current step by the number of organisms found in the previous step (Equation 2.2).

$$\text{stepwise percent recovery} = \frac{\text{total number of organisms in sample}}{\text{total number of organisms in previous step}} \times 100 \qquad (2.2)$$

These equations helped establish the efficiency of recoveries between different steps used in the entire process.

DOWNSTREAM CONCENTRATION WITH ADSORPTIVE FIBERS

The powder fiber matrix (NanoCeram powder; Argonide) was tested for its capacity to adsorb viral particles and to optimize the elution of bound viral particles into a much smaller volume. Initial experiments were performed in a 10-mL volume of dechlorinated tap water with 0.5 g of fiber. Additional experiments were conducted with larger amounts of fiber and larger volumes of tap water. Initial experiments were performed with bacteriophage PP7 and then expanded to phage MS2 and poliovirus. (Details of the adsorbent fiber spiked challenge are given in appendix E.)

Dechlorinated tap water (13 mL) (free chlorine concentration should be below 0.01 mg/L) was added to a 15-mL centrifuge tube, and ~5,000 pfu of bacteriophage PP7 from a stock solution was added to it and mixed. Three milliliters of the suspension was then collected for determination of the initial phage concentration. The powder (0.5 g) was then added to the remaining 10 mL of tap water and placed on a shaker (orbital) for 15 minutes at 4°C. The tube was then removed from the shaker and centrifuged at 750× *g* for 15 minutes to pellet the fibers. Finally, the supernatant was collected and stored for plaque assay.

The pellet was resuspended in an elution solution (10 mL). The elution solution may have been pH adjusted with 1*M* NaOH or 1*M* HCl, depending on the target pH and the particular experiment. Next, the mixture was placed on the shaker for 15 minutes at 4°C. Elution solutions tested included bovine serum albumin (BSA), calf serum, fetal calf serum, beef extract, glycine, and NaPP (Table 2.2). The tube was then centrifuged and the supernatant and pellet were stored for analyses by plaque assay. In most experiments, a secondary elution was performed to improve

Table 2.2
List of agents tested for elution of phage PP7 from adsorptive fibers

Agent	Concentration	pH
BSA	5%, 10%, 15%	5.0, 9.3, 10.0, and 11.0
Beef extract and glycine	1.5%, 0.25%	9.3
Calf serum	10% and 15%	9.3
Calf serum + Tween 80	5% and 0.1%	9.3
Glycine	0.05*M*	9.3
Glycine and Tween 80	0.05*M* and 0.1%	5.0 and 9.3
Minimum essential media (MEM) + 10% fetal bovine serum	1 × MEM	9.3 and 10.0
Fetal bovine serum	5%, 10%, and 15%	9.3

recovery and recovery consistency. The phage or virus concentration in the eluent and pellet were determined by plaque assay in a similar manner as experiments that used a single elution step.

A limited number of experiments were also conducted with concentrates from a 10-L sample of tap water. The concentrate contained Tween 80 and 0.05*M* glycine at pH 7.0 from the elution step. The amount of powder added to each experiment varied depending on the volume of concentrate and the desired powder concentration. The largest volume tested was 200 mL of concentrate.

IMT Ultrafilters

Polysulfone 20,000–80,000 Da MWCO hollow fiber ultrafilters (IMT) were used in 10-L tap water experiments. The hollow fiber ultrafilter consists of hollow tubes of filter material bundled together. The flow of the filtered water passed from the inside to the outside of the fiber. The fiber diameter was 0.4 mm. The filters were tested in two configurations, single-bore and multi-bore. The multi-bore had six hollow tubes housed within a single fiber. The single-bore consisted of a single hollow tube within a fiber. Table 2.3 compares all configurations of the filters used in this study.

Filtration Procedure

The hollow fiber ultrafilter module was fitted into a filtration system (pump head 77250-62, Cole-Parmer) (Figure 2.1). Before and after each experiment, the ultrafilter module was sanitized by circulating a solution containing 100 mg/L free sodium hypochlorite for 30 minutes. The free sodium hypochlorite concentration was determined at the end of sanitation by measuring the absorbance at 530 nm by the DPD method (DR/2010, Hach) The ultrafilter module was flushed with sterile ultrapure water (UPW) until the residual concentration of free sodium hypochlorite reached <0.04 mg/L.

Table 2.3
Comparison of the hollow fiber ultrafilters used in this study

Ultrafilter	Filter type	Membrane	Pore size MWCO (Da)	Surface filtration Area, ft^2 (m^2)		Volume (L)
Microza hollow fiber	Pencil	Polyacrylonitrile	50,000	0.18	(0.017)	2
	Lab	Polyacrylonitrile	50,000	2.2	(0.2)	10
	Pilot	Polyacrylonitrile	50,000	11	(1)	100
IMT	Small	Polysulfone	50,000	5.5	(0.5)	10

MWCO = molecular weight cutoff.

Viruses were suspended in 5-L volumes of water buffered with PBS (pH 7.2, 1.54 m*M* KH_2PO_4 [potassium phosphate], 154 m*M* NaCl [sodium chloride], 6.05 m*M* Na_2PO_4 [sodium phosphate]) with a virus concentration of ~1,000 pfu/mL. A 10× solution of PBS was added to the 5-L sample prior to the addition of a virus. Viruses were resuspended in ultrapure, tap, ground-, or surface water buffered with PBS. The virus solution was stirred for 10 minutes, then recirculated throughout the UF unit by crossflow circulation for 5 minutes.

The samples were processed at 15 psi. Each experiment was designed such that the retentate was recirculated to a central reservoir (Figure 2.1). Filtration was terminated when only the holdup volume (volume of fluid contained in the filter apparatus) remained. For each experiment, the virus concentration was determined for the initial virus suspension, retentate, and overall "bulk" permeate by plaque assay.

Blocking and Elution Methods

The membrane was blocked with proteinaceous agents prior to filtration of the virus suspension. This was accomplished with a 100-mL suspension of blocking solution. The solution was allowed to flow across the membrane (crossflow) for 1 hour (no permeate flow) or blocked overnight with agitation at room temperature. All blocking solutions were prefiltered through a 47-mm nylon, 66-membrane filter (0.2-µm; NT, Pall Corp.) prior to use. Unbound blocking material was removed by flushing the system with 500 mL of UPW (single pass) prior to filtration. If NaPP was used, the solution was recirculated through the filter for 1 hour prior to use.

Filtration was performed at a permeate flow rate of 400 mL/min. The pressure was less than 10 psi. Elution of viral particles from particulates or the filter was completed after the concentration step by the addition of 0.05*M* glycine (final concentration) at pH 7.0 to the retentate. The retentate (eluent added) was then passed through the membrane for 30 minutes in the crossflow configuration only (no permeate flow). A sample of the retentate was then taken for analysis. Several blocking and elution agents were tested (Table 2.4; Figure 2.6).

Table 2.4
Agents used to reduce viral adsorption to the filter or to resuspend bound virus

Blocking agents	Added to virus suspension	Elution
5% calf serum (Gibco-BRL; Grand Island, N.Y.)	0.1% Tween 80	0.05*M* glycine pH 7.0, 9.0 Bovine serum albumin, calf serum, and other elution agents

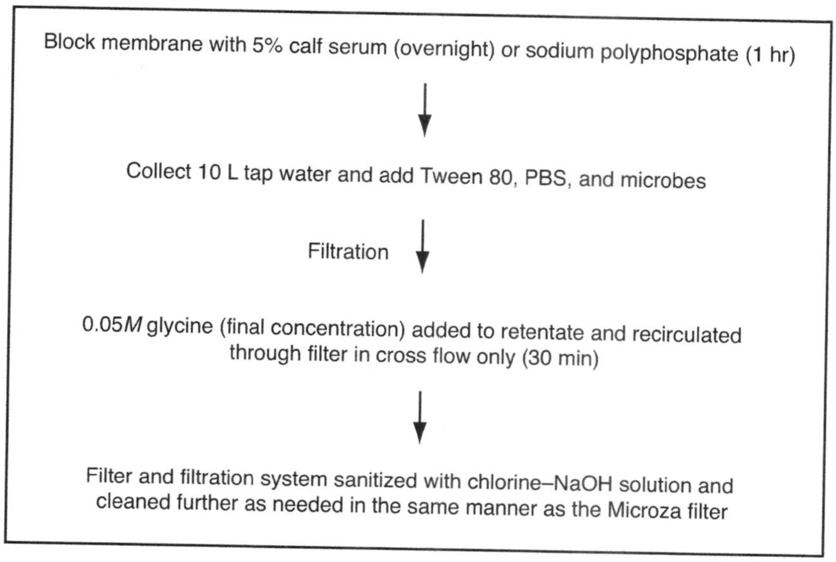

Figure 2.6 Optimized filtration scheme (10 L) for IMT hollow fiber UF system

Backflush

For some experiments, after filtration was completed, a solution containing 0.05*M* glycine (pH 7.0), 0.1% Tween 80 in PBS, was filtered in the reverse direction. Depending on the experiment, 50–75 mL of the elution solution was filtered in the reverse direction, and the entire volume was collected and sampled to determine virus concentration. Pressure during the backflush did not exceed 25 psi. A backflush sample was then collected for the viral plaque assay or pooled with the retentate for the oocyst assay. The backflush was accomplished by connecting the peristaltic pump tubing to the permeate port on the filter module and forcing 50 mL of eluent through to the inside of the hollow fiber (<25 psi). The backflush step was completed after the retentate was removed. In some experiments, the retentate with added eluent was recirculated in a crossflow pattern for 30 minutes after backflushing. In other experiments, multiple backflush steps were attempted to increase the recovery of bound viruses.

CHAPTER 3
RESULTS AND DISCUSSION

ULTRAFILTRATION OF 10 L OF TAP WATER

Recovery experiments included phage PP7, poliovirus, *E. coli* K-12, *B. globigii* spores and vegetative cells, and *Cryptosporidium* oocysts (Table 3.1). Mean recoveries of the model organisms of 25.1%–78.8% were observed when filters were used immediately after the blocking step. Although recoveries of *Cryptosporidium* oocysts were the lowest, they represent the combined effects of filtration, centrifugation, and IMS. Recovery that represented only the UF process was not measured.

Recoveries with *Bacillus* spores were similar to those of *E. coli*. Treatment of the spore stock with lysozyme greatly reduced the amount of vegetative cells present in the stock preparation as well as the clumping of spores and vegetative cells as visualized by microscopy. (The results are not shown in this report.)

Other experiments with UF involved the addition of thimerasol in the blocking solution. The rationale for use of thimerasol in the blocking solution was to provide protection against bacterial contamination during long-term storage (2–4 weeks) such that the blocked modules could be stored for longer periods and then used without further preparation. It would be advantageous to be able to use the blocked filters immediately by rinsing the blocking solution from the modules and then begin processing the water sample.

Storage of Modules

Recovery of test organisms was determined for Microza ultrafilters that were blocked with calf serum and then stored for 2 and 4 weeks before use. The blocked filters were briefly rinsed and then immediately used under simulated conditions of an actual event. In such a situation, it is desirable for processing of the samples to begin immediately upon arrival. Two types of experiments were conducted: filters blocked and stored in DI water, and those blocked and stored in 0.05% calf serum with 0.01% thimerasol.

During storage, plate counts of the storage media were performed to determine whether bacterial growth was occurring. The results indicate that storage of the filter units for up to 1 month did not cause measurable growth of bacteria within the units for filters stored in either sterile DI water or blocking solution containing thimerasol. The effects of storage of the blocked filters on concentration efficiency was evaluated by the recovery of phage PP7, poliovirus, *E. coli* K-12, and *Bacillus* spores from filters stored for 2 and 4 weeks as compared to unstored filters. Recoveries did not appear to be adversely affected from storage and use of blocked filters for 2 or 4 weeks (Tables 3.2 and 3.3). These results indicate that the storage of blocked filters is feasible. In addition, although thimerasol can be used to control bacterial growth, if necessary, bacterial contamination was not a problem even when thimerasol was not used. The effectiveness of thimerasol for controlling bacterial growth was further demonstrated by the addition of raw sewage (1,000 cfu/mL of naturally occurring sewage bacteria in 200 mL of storage solution with 0.01% thimerasol) to the filters. No bacterial colonies were detected in any of the stored filters (0.1-mL sample volume).

Table 3.1
Mean recoveries and standard deviations of test organisms with the hollow fiber ultrafilter for 10 L of tap water*

Organism	Number of replicates	Mean recovery, % (SD)
Phage PP7	22	77.8 (30.1)
Poliovirus	5	74.7 (28.7)
E. coli K-12	6	76.8 (40.4)
B. globigii vegetative cells and spores	5	69.1 (21.4)
B. globigii spores	6	72.3 (13.4)
Cryptosporidium oocysts	10	25.1 (14.2)

* Filters were blocked and used immediately.
SD = standard deviation.

Growth characteristics of selected bacterial agents were also measured in the presence of varying concentrations of thimerasol (Table 3.4 and Figure 3.1). These results were then used to select a thimerasol concentration to be used in the storage experiments. Some variation in growth inhibition was noted. *B. globigii* vegetative cells and spores were the most resistant. From these results, a 0.01% concentration of thimerasol was selected as a preservative.

100-L MICROZA ULTRAFILTRATION

Recovery From Initial Ultrafiltration Step

The same filtration conditions that were done with 10 L were then tested with 100 L using a larger surface area pilot filter module (Flow Solutions, Austin, Texas; Table 2.3). Results are presented in Table 3.5. Filtration of 100 L (to 1 L) was quite rapid, occurring in ~24 minutes (~5,000 mL/min).

A modification to the basic procedure was also tested where 0.01% NaPP was added to the initial sample and limited tests were conducted where bovine serum was added as an amendment. The results indicate that the addition of NaPP improved the recovery of some of the organisms tested (Table 3.6). Viral recovery without NaPP appeared to be lower than in the 10-L system.

The addition of 0.01% NaPP increased the recovery of most agents in the initial retentate (Table 3.3). Average recoveries of phages MS2 and PP7, and *E. coli* K-12 increased by 14.0%, 9.4%, and 15.1%, respectively. Poliovirus decreased by 1.9%; however, the standard deviation of this virus markedly decreased with the NaPP amendment strategy.

The addition of 0.01% calf serum instead of NaPP to the initial 100-L sample also increased the average recovery of phage PP7 in the initial retentate sample when compared with the original protocol. However, the presence of calf serum in the final concentrate interfered with the use of adsorptive fibers for downstream viral processes (data not provided). Presumably, the presence of bovine serum (a known eluent) in the final concentrate prohibited binding between suspended viruses and the adsorptive fibers. There were also concerns about decreased filterability during the secondary UF process.

Table 3.2
Recovery of selected agents from stored (with and without thimerasol for 2 and 4 weeks) Microza hollow fiber ultrafilters (10 L)

Organism	Storage duration	Storage condition	Contamination	Average and SD*
Phage PP7	No storage	NA	NA	77.8 (22), SD 30.1
Phage PP7	2 weeks	Sterile water	None	75.16 (6), SD 30.6
Phage PP7	2 weeks	0.01% thimerasol block	None	74.6 (5), SD 19.3
E. coli K-12	No storage	NA	NA	76.8 (5), SD 40.4
E. coli K-12	2 weeks	Sterile water	None	68.3 (6), SD 31
E. coli K-12	2 weeks	0.01% thimerasol block	NA	NA
Bacillus spores	No storage	NA	NA	69.1 (5), SD 13.4
Bacillus spores	2 weeks	Sterile water	None	129 (4), SD 47.64
Bacillus spores	2 weeks	0.01% thimerasol block	None	103 (3), SD 5.1
Poliovirus	No storage	NA	NA	74.7 (5), SD 28.7
Poliovirus	2 weeks	Sterile water	None	53.5 (3), SD 12.01
Poliovirus	2 weeks	0.01% thimerasol block	None	62.7 (4), SD 30.96
Phage PP7	4 weeks	Sterile water	None	57.6 (5), SD 20.84
Phage PP7	4 weeks	0.01% thimerasol block	None	70 (4), SD 33
E. coli K-12	4 weeks	Sterile water	None	78.3 (3), SD 27.84
E. coli K-12	4 weeks	0.01% thimerasol block	None	46 (2), SD 14
Poliovirus	4 weeks	Sterile water	None	48 (3), SD 12.77
Poliovirus	4 weeks	0.01% thimerasol block	None	96.3 (3), SD 13.0
Bacillus spores	4 weeks	Sterile water	None	86.6 (3), SD 11.7
Bacillus spores	4 weeks	0.01% thimerasol block	None	82.2 (4), SD 22

* The number of replicates is shown in parentheses. Recoveries are from UF and after centrifugation of the concentrate at 2,600× *g*.
SD = standard deviation; NA = not analyzed.

Downstream Processes to Further Concentrate Viral Particles

Two downstream concentration methods were tested (secondary UF process and adsorptive fibers) and found to be effective. Mean recoveries of >50% final recovery were observed for all organisms except *C. parvum* oocysts. The downstream processes were conducted with concentrates from 100-L samples that were concentrated with Microza hollow fiber modules.

Recovery With Secondary Ultrafiltration Process

Recovery of viral agents after the initial UF step for 100 L of drinking water was tested using several methods. One of the most promising methods is presented in Table 3.6. A secondary UF step was used after the initial concentrate was centrifuged to pellet cellular organisms and particulates. After the secondary UF process, the concentrate was further reduced in volume using

Table 3.3
Comparison of three 100-L UF sample amendment strategies: Average recoveries of organisms in primary concentrate*

Organism	Standard method: 10 L 10× PBS 0.01% Tween 80, % (SD)	NaPP method:† 0.01% NaPP 10 L 10× PBS 0.01% Tween 80, % (SD)	Calf serum method:‡ 0.01% calf serum 10 L 10× PBS 0.01% Tween 80, % (SD)
Phage MS2	50.1 (15.2), n = 6	74.1 (9.3), n = 5	
Phage PP7	52.8 (11.4), n = 13	62.2 (6.9), n = 8	70.9 (11.1), n = 3
Poliovirus	66.7 (22.7), n = 4	64.8 (11.5), n = 3	
B. globigii	None	87.5 (15.3), n = 3	
E. coli K-12	84.8 (10.6), n = 3	98.9 (9.3), n = 5	

*All initial water samples were concentrated to ~1 L, which were then amended to a final concentration of 0.05*M* glycine and recirculated for 30 minutes, yielding the primary concentrate. Filters were blocked with 0.05% calf serum prior to use.
†NaPP (0.01%) was added to the initial 100-L suspension together with PBS and Tween 80.
‡Calf serum (0.01%) was added to the initial 100-L suspension together with PBS and Tween 80. The use of calf serum as an amendment resulted in a slow flow rate; therefore, only phage PP7 was used.
SD = standard deviation.

Table 3.4
Cumulative minimum inhibitory concentration* of thimerasol (0.1–0.000001%)

Organism	0.1	0.01	0.001	0.0001	0.00001	0.000001	Control
E. coli XL1B	0.023	0.034	0.023	0.598	0.64	Not analyzed	0.60
B. globigii	0.016	0.008	0.011	0.009	0.30	0.40	0.40
E. coli K-12	0.018	0.011	0.013	0.520	0.56	0.51	0.51
B. globigii spores	0.117	0.103	0.114	0.113	0.42	0.61	0.61

*Average absorbances (620 nm) for each bacteria (n = 2) after 17 hours of culture in growth medium.

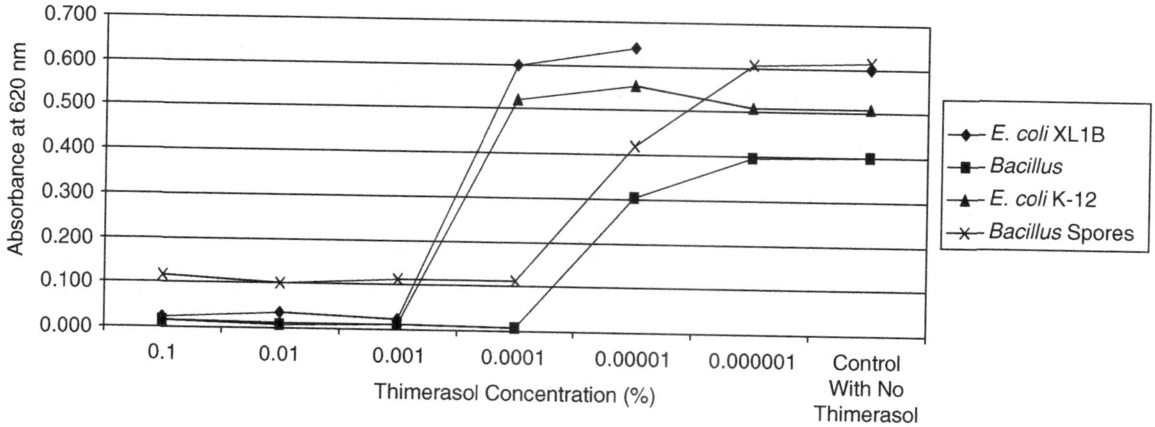

Figure 3.1 Cumulative minimum inhibitory concentration of thimerasol for selected microbes (n = 2)

Table 3.5
Percent recoveries of poliovirus and phage PP7 using adsorptive powder

Organism	Powder concentration	% Adsorbed	% Elution 1*	% Elution 2†	% Recovery‡	% Recovery from elution§
Phage PP7	1 g	4	28	6.2	11.4	34.2
	2 g	77.6	24.1	11	1.6	35.1
	2 g	62.5	20.8	7.4	5.8	28.2
	4 g	98.6	81.2	28	0.7	**109.2**
	4 g	99.13	45	21	4.5	**66.0**
	4 g	98.5	41	13	4.1	**54.0**
Poliovirus	1 g	31	24.7	4.5	1.8	29.2
	2 g	79.1	17.4	4.7	6.5	22.1
	4 g	98.2	92.5	14.9	11.2	**107.4**
	4 g	92.7	60.9	31.8	6.2	**92.7**
	4 g	98.3	74.2	7.5	4.5	**81.7**

NOTE: The 200-mL concentrates are from a 10-L UF process of tap water. Elution performed with Dulbeccos minimum essential media with 10% fetal bovine or calf serum (DMEM-10) at pH 9.3. Bold numbers represent the conditions that worked best.

*Percentage recovered after the first centrifugation.

†Percentage recovered after the second centrifugation.

‡Percentage recovered from the pellet after the second centrifugation.

§Percentage recovered from all samples.

Table 3.6
Average recovery of organisms from 100 L of drinking water using a two-step UF protocol and NaPP amendment*

Agent	Average recovery	Retentate from first UF step, % (SD)	Postcentrifugation supernatant, % (SD)	Postcentrifugation pellet, % (SD)	Retentate from secondary UF step, % (SD)	Post-spin columns, % (SD)
Phage PP7, n = 3	Total	62.2 (1.8)	63.0 (2.7)	NA	55.4 (15.6)	**55.2** (11.8)
	Stepwise	62.2 (1.8)	101.3 (3.0)	NA	87.5 (22.9)	101.3 (10.1)
Phage MS2, n = 3	Total	74.1 (11.8)	67.7 (17.4)	NA	55.0 (8.9)	**72.3** (11.8)
	Stepwise	74.1 (11.8)	90.5 (8.4)	NA	83.8 (22.4)	136.1 (46.8)
Poliovirus 2, n = 3	Total	64.8 (11.5)	NA	NA	81.8 (38.2)	**85.8** (20.2)
	Stepwise	64.8 (11.5)	NA	NA	123.5 (40.7)	112.5 (25.8)
Bacillus spores, n = 3	Total	87.5 (15.3)	7.4 (2.2)	**54.1** (7.0)	NA	NA
	Stepwise	87.5 (15.3)	9.0 (4.6)	63.6 (16.2)	NA	NA
E. coli K-12	Total	98.9 (9.3) n = 5	37.2 n = 2	**50.2** (19.6) n = 5	NA	NA
	Stepwise	98.9 (9.3)	40.0 n = 2	50.4 (19.5) n = 5	NA	NA
C. parvum, n = 3	Total	NA	NA	**25.9** (6.5)	NA	NA
	Stepwise	NA	NA	NA	NA	NA

*Large-scale UF with 0.01% NaPP, 0.01% Tween, 1X PBS added to the 100-L volume, and a 30-minute recirculation with 10% 0.5*M* glycine (0.05*M* in sample) of the concentrate. Concentrate was centrifuged at 2,600X *g* for 20 minutes. Supernatant was further concentrated with pencil module UF 50,000 Da MWCO to ~50 mL. Spin columns were used to reduce volume to 10–15 mL (usually two spins at 3,000 rpm for 10 minutes).

‡Percentage recoveries in the final concentrate (recovery after the entire process) appear in bold.

NA = not analyzed; SD = standard deviation.

30,000 Da MWCO spin columns. The entire process for viral concentration was less than 3 hours and reduced the volume from 100 L to 10–15 mL. Recoveries for each major step as well as the cumulative recoveries were recorded.

For all viruses tested, >55% was recovered from the final concentrate (cumulative recovery Table 3.6). The lowest stepwise recoveries for all viruses were observed in the initial large-scale UF.

Stepwise recoveries of phages PP7 and MS2 and poliovirus after the spin column were greater than 100%, which resulted in final cumulative recoveries that were higher than some of the intermediate recovery data points.

The NaPP amendment strategy resulted in the recovery of *E. coli* K-12 and *B. globigii* spores of 98.8% and 87.5%, respectively, after the initial UF process. The subsequent centrifugation step demonstrated less efficient stepwise recoveries of 50.4% and 63.6% for *E. coli* and *B. globigii*. Recovery of *E. coli* K-12 in the supernatant was 40.0% (stepwise). When added to the recovery of the pellet, 90.4% of *E. coli* K-12 from the previous step was accounted for. Only 9.0% of *B. globigii* spores from the previous step were recovered from the supernatant, yielding an average stepwise recovery of 72.6% when the pellet and supernatant recoveries were combined.

Centrifugation recovery efficiencies for *E. coli* and *B. globigii* spores decreased the final cumulative recoveries of these bacteria to 50.2% and 54.1%, respectively (Table 3.6).

Cryptosporidium concentrations were not tested in the intermediary steps of the concentration process because of the low concentration of oocysts that were added to the initial sample. The average final recovery of *C. parvum* oocysts in the final pellet was 25.9%.

The Effects of Amendments and Elution Agents

Tween 80 is a nonionic surfactant that can reduce the hydrophobic interactions between microbes and surfaces such as the filter or particulates (Sharma, Chang, and Yen 1985) and has been used in previous concentration studies to reduce hydrophobic interactions (Hill et al. 2005; Morales-Morales et al. 2003; Sharma, Chang, and Yen 1985). The addition of Tween 80 as part of the initial amendment was to reduce this interaction from the initial filtration process. It was also added in the concentrate prior to the secondary UF process (part of the two-step UF process for 100 L of tap water).

Polyphosphates are highly negatively charged chemicals that have been used as a dispersant in soil and for preventing aggregation of microbial agents (Lepesteur, Blasdall, and Ashbolt 2003; Papo, Piani, and Ricceri 2002). Recently, NaPP was used as an amendment for recovery of microbes from large volumes of water (Hill et al. 2005).

Recovery of Viral Agents With Adsorptive Fibers

Several elution constituents were evaluated to elute viral agents from a positively charged fiber. Conditions tested included 5% and 10% calf serum; 1.5% beef extract; 0.1% Tween 80; glycine; 5%, 10%, and 15% BSA; 10% calf serum plus 0.1% Tween 80; and DMEM-10. Different volumes of tap water (unconcentrated and concentrated tap water samples) and amounts of fiber were used (data not provided).

Initial recoveries were performed with unconcentrated tap water to determine the binding efficiency and overall recovery. Variables tested included varying the elution agent, pH, amount of fiber added, and the volumes of water (data not provided). The results of these experiments clearly demonstrated that the fiber was highly efficient at adsorption. Recoveries were highest with DMEM-10 as the elution agent.

When the fiber was added to concentrates from 10-L samples, recoveries were most efficient when DMEM-10 was used as an eluent and 4 g of fiber was added to 200 mL of concentrate (Table 3.5). The concentration of fiber was later increased to 6 g/200 mL (Tables 3.7 and 3.8).

Results from 10-L concentrates indicate that the final concentrate volume after elution was 10–20 mL. The results also indicate that phage PP7 and poliovirus were efficiently recovered when DMEM-10 and 0.01% NaPP was used as the eluent (Tables 3.7–3.9). The results clearly indicate that the fibers were still highly effective at binding viral particles even in the presence of Tween 80 and glycine. This attribute is highly desired since both were present after the initial UF step.

Table 3.7
Optimization of three viral concentration strategies using adsorptive fibers:[*]
Average recoveries of phage PP7

Variables tested[†] (eluent strategies using 6 g of adsorptive fibers)	% Adsorbed to fibers (SD)	% Recovery after elution[‡] (SD)	% Recovery from pellet[§] (SD)
0.01% NaPP (sodium polyphosphate)	97.3%	58.2%	4.1%
DMEM-10	96.48 (2.3)	60.2% (4.2)	16.9% (2.0)
DMEM-10 with 0.01% NaPP	96.48 (2.3)	70.3% (3.7)	10.9% (1.1)

[*]Optimization was performed using 200-mL concentrates.
[†]n = 3.
[‡]DMEM-10 (pH 9.3) with 10% fetal bovine serum.
[§]Recovery from pellet portion after centrifugation (postelution).
SD = standard deviation.

Table 3.8
Average recovery of viral agents during downstream concentration processes (adsorptive fibers and spin columns)[*]

Virus	Recovery type	% Bound to fibers (SD)	% Recovery after elution[†] (SD)	% Recovery from the pellet (SD)	% Recovery after spin column (SD)
Phage MS2	Cumulative	97.7 (1.6), n = 3	72.5 (5.2)	3.9 (1.3)	76.5 (4.3)
	Stepwise	97.7 (1.6), n = 3	92.1 (5.7)	7.2 (1.7)	108.1 (5.8)
Phage PP7	Cumulative	97.2 (1.8), n = 3	70.7 (2.4)	8.1 (0.8)	60.5 (1.2)
	Stepwise	97.2 (1.8), n = 3	72.6 (2.17)	11.4 (0.6)	86.1 (2.3)
Poliovirus	Cumulative	96.9 (1.7), n = 2	76.5 (5.1)	6.7 (2.2)	83.4 (5.2)
	Stepwise	96.9 (1.7), n = 2	79.9 (5.5)	7.5 (1.9)	110.1 (1.7)

[*]All 100-L samples were concentrated to ~1 L, amended to a 0.05*M* concentration of glycine, and recirculated for 30 minutes, yielding primary concentrates. Primary concentrates were centrifuged to enumerate cellular organisms. The supernatant was treated with adsorptive fibers, reducing the volume to ~200 mL (6 g fiber/200 mL of concentrate). Spin columns were used to reduce the final volume to ~15 mL.
†Viral agents were eluted with 0.01 NaPP and DMEM with 10% fetal bovine serum.
SD = standard deviation.

Table 3.9
Comparison of two 1-L viral downstream concentration processes*

Organism	Secondary UF method coupled with spin columns,† % (SD)	Adsorptive fiber method coupled with spin columns,‡ % (SD)
Phage MS2	141.1 (20.4), n = 4	76.5 (4.3), n = 3
Phage PP7	88.6 (8.6), n = 5	60.6 (1.6), n = 3
Poliovirus (Sabin strain type 2)	138.9 (20.2), n = 3	83.4 (5.2), n = 2

*Recoveries reflect percentages that do not include the initial UF and centrifugation steps.
†The supernatant was further concentrated with a small-scale ultrafilter to ~50 mL. Spin columns were then used to reduce the final volume to ~15 mL.
‡The supernatant was treated with adsorptive fibers and eluted with DMEM-10 (pH 9.3) and 0.01% NaPP solution. Spin columns were used to reduce the final volume to ~15 mL.
SD = standard deviation.

Comparison of Secondary Ultrafiltration and Adsorptive Fiber for Downstream Sample Concentration

Two downstream concentration processes were compared for recovery efficiency. Recoveries of three viral agents were determined with a secondary UF process versus the use of the adsorptive fiber. Each was further concentrated with spin columns. Both approaches produced efficient recoveries (Table 3.9). These experiments were performed starting with 1-L concentrates from a 100-L UF process (Tween 80 and glycine were present in the concentrate).

Recovery From 100-L Initial Volume to Final Concentrate

For all viruses tested, >55% of the viruses was recovered from the final concentrate (Table 3.5) when the two UF processes were completed in <3 hours. The lowest stepwise recoveries for all viruses were observed in the initial large-scale UF.

Results with the adsorptive fiber approach suggest that the concentration of 100 L to 10–20 mL can be accomplished within 4 hours. The process includes UF, centrifugation of the concentrate to pellet cellular organisms, and then the further concentration of the viral agents by adsorptive fibers and spin columns.

The adsorptive fibers were then tested with other conditions in the 200-mL concentrates from 100 L of tap water. The addition of NaPP with DMEM-10 in the eluent appeared to enhance recovery (Table 3.10). The concentration of phage PP7 in the pellet was also examined to assess the effectiveness of the elution process.

Use of 0.01% NaPP, 0.01% Tween 80, and 10 L 10× PBS yielded optimal simultaneous recovery of viruses and bacteria in the primary UF step using the large-scale ultrafilter module. Except for poliovirus type 2, the average recovery efficiency of every organism sampled increased when NaPP was used (Table 3.3). The average recovery of poliovirus decreased slightly; however, the standard deviation in the data set using NaPP was much smaller than that of the original protocol. The smaller variation in recoveries may be due to enhanced recovery over a range of water qualities when NaPP was utilized. Experiments using the NaPP sample amendment strategy

Table 3.10
Recovery of phage PP7 and *E. coli* with a 90-mm adsorptive fiber filter using different elution conditions

Eluent	Eluent pH	Organism	Input*	% Adsorption	% Recovery†	Flow (mL/min)	Filtration time (min)	Pressure (psi)
1.5% BE + 0.25% glycine	9.3	Phage PP7	4.5×10^5	100	9.2	430	21	4
10% Na_2CO_3 (sodium carbonate)	9.33	Phage PP7	3.0×10^5	100	0.27	430	30	5
10% calf serum	9.73	Phage PP7	1.8×10^7	100	0.5	530	19	5
5% BSA	Not measured	Phage PP7	1.8×10^7	96.3	2.7	1,000	10	10
10% BSA	Not measured	Phage PP7	2.0×10^7	100	0.8	420	18	5
DMEM-10	9.28	Phage PP7	2.0×10^5	100	34	298	34	4
DMEM-10	10.02	Phage PP7	1.9×10^7	100	TNTC	300	29	3
DMEM-10	10	Phage PP7	1.4×10^7	100	62	230	40	4
DMEM-10	10.09	Phage PP7	1.2×10^7	100	3.3	290	42	3
DMEM-10	10	Phage PP7	1.6×10^7	100	4.8	310	30	2
		E. coli	8.2×10^5	100	0.3	310	30	2
DMEM-10	10.05	Phage PP7	1.4×10^7	99.64	2.1	200	44	2
		E. coli	1.0×10^6	100	2.6			
DMEM-10	9.99	Phage PP7	1.0×10^7	100	6.5	310	30	3
		E. coli	Contaminant	ND	ND			

*The total plaque-forming units in the initial 10-L suspension.
†Percent recovered in the eluent.
BE = beef extract; BSA = bovine serum albumen; ND = no data; TNTC = too numerous to count.

without a glycine recirculation step showed little to no improvement over the original protocol for phage PP7 (not given).

The 3-hour timeline for this method consisted of 1.5 hours for the initial UF process that included the recirculation of the sample (10 minutes), UF concentration (~20 minutes), recirculation/elution (30 minutes), and centrifugation (20 minutes). After centrifugation, the pellet was ready for processing. Further concentration for viral analysis required 1 hour and involved mixing of the concentrate (5 minutes), secondary UF (~35 minutes), and spin column processing (~20 minutes) (Figure 2.5). An additional 30-minute margin was added to the timeline to adjust for time required between steps.

Adsorptive Fiber 90-mm Disks

Ninety-millimeter disks of the adsorptive fibers attached to a filter backing were used to bind and elute viral agents from 10 L of tap water (appendix F). Experiments have shown efficient adsorption of phage PP7 but low elution efficiency. Recoveries from tap water were less than 10% (Table 3.10). Because of low recoveries, no additional testing was performed.

Table 3.11
Recoveries from 10 L using IMT filters*

	Phage PP7	Poliovirus	*Bacillus* spores	*E. coli* K-12
Average (standard deviation)	58.3% n = 16	66.7% n = 4	53.6% n = 3	66.3% n = 3
Standard deviation	20.4	22.7	6.8	17.6

*An earlier prototype version of the IMT filter. Filters were blocked with 0.05% calf serum.

Table 3.12
Comparison of average percent recoveries of organisms in the concentrate between Microza and IMT filter modules using various types of blocking strategies*

Organism	Microza ultrafilter,† % (SD)	IMT filter with bovine serum as the blocking agent, % (SD)	IMT filter with NaPP as the blocking agent,‡ % (SD)
Phage MS2	NA	NA	77.0 (32.2), n = 3
Phage PP7	77.8 (30.1), n = 22	59.0 (30.2), n = 3	63.0 (24.2), n = 5
Poliovirus (type 2)	74.7 (28.7), n = 5	NA	68.5 (20.2), n = 3
B. globigii spores	76.7 (52.6), n = 3	72.8, n = 1	86.7 (9.60), n = 3
E. coli K-12	73.8 (33.3), n = 8	NA	83.0 (3.00), n = 3
C. parvum oocysts§	37.5 (10.9), n = 10	NA	39.8 (3.00), n = 3

*All 10-L tap water samples were concentrated to ~200 mL, amended with 0.05*M* glycine, and recirculated for 30 minutes.
†Microza ultrafilters were blocked using 5% bovine serum.
‡IMT filters were blocked using 0.1% NaPP.
§Recoveries are included potential losses that occurred during the filtration and IMS and IFA processes.
NA = not analyzed; SD = standard deviation.

IMT Ultrafilters

Lower recoveries were observed when 0.01% Tween 80 and PBS were added to the initial sample. Elution was achieved with the addition of glycine (0.05*M*) in the concentrate which was recirculated in the crossflow mode for 30 minutes. The standard method used in the Microza filters that was also applied to the IMT filters (Table 3.11). Recoveries for viral agents were improved with the addition of NaPP instead of calf serum in the initial sample (Table 3.12).

Although in other ultrafilters, the use of a backflush process after filtration has been found to be a useful procedure to increase recoveries (Hill et al. 2005), a backflush process did not appear to increase recoveries and was not pursued further.

Recoveries between Microza and IMT filters are very similar for 10-L samples when the polyphosphate amendment was used (Table 3.12). These results indicate that IMT filters may be a good alternative to Microza ultrafilters. Larger 100-L prototype modules with 3.8 m^2 of surface area have been designed and will be tested in the future.

SUMMARY

This study developed an integrated approach for the rapid and efficient concentration of viral, bacterial, and protozoan agents from large volumes of drinking water. The method allowed for >50% recovery of viral and bacterial agents, and >25% recovery of protozoan agents from 100 L of drinking water in ~3 hours.

Two ultrafilters were evaluated as well as several downstream concentration processes in order to develop a complete method. The overall process involved an initial UF process to concentrate 100 L of drinking water to a 1-L concentrate. Two downstream processes were then evaluated to further concentrate viral agents from primary concentrates produced by large-scale UF. One process included the use of a secondary UF process coupled with spin columns, whereas the alternative process included the use of positively charged adsorptive fibers coupled with spin columns.

Bacteriophage PP7 was used as a control for most experiments. The small size and ease of assay made phage PP7 a convenient system to use for monitoring filter integrity and the overall concentration process.

Initial Ultrafiltration

Previous studies have used UF processes in various volumes of tap, ground-, river, and sea waters. Table 3.13 illustrates a representative sampling of previous research using UF for the initial concentration of water. Recoveries described by the literature range from 22% to 148%, 47% to 92%, and 14% to 65% for viruses, bacteria, and protozoa, respectively. In a study that processed large volumes of water, Bicknell, Dziewulski, and Sturman (1985) used UF to reduce 100 L of tap water to retentate volumes between 100 and 750 mL (depending on backflush volumes) in <3 hours. The 100-L volume used in this study was one of the larger volumes that has been tested for UF systems, and the initial concentration process could be completed in ~1.5 hours (including the centrifugation process).

This study confirms the utility of UF to efficiently concentrate a wide variety of microbes in a time frame that can unify the process into a single method appropriate for all types of pathogens. The robustness of the Microza ultrafilters to concentrate microbes from a wide variety of water qualities has been further confirmed by previous studies with surface and groundwater samples (Morales-Morales et al. 2003; Olszewski, Winona, and Oshima 2001, 2005).

The use of NaPP in the initial 100-L suspension increased recoveries in the concentrates for viral, bacterial, and protozoan agents (Table 3.3). The effectiveness of NaPP in the concentration process has recently been reported by Hill et al. 2005. The use of a calf serum amendment also increased recoveries in primary concentrates for phage PP7; however, this strategy may interfere with further downstream processes such as the secondary UF step, adsorption to the positively charged fibers, or the use of spin columns. The effect of NaPP on the survival of bacterial agents should be investigated before its formal inclusion in any concentration method because of its potential to affect microbial viability (Knabel, Walker, and Hartman 1991). However, no adverse effects were noted in this study.

Table 3.13
Comparison of UF techniques as a first-step concentration process

Type of water	Initial volume (L)	Ultrafilter type (MWCO, Da)	Time (hr)	Retentate volume	Recovery of seeded agents	Reference
Ground	400	UF	~3	~10 L	22% poliovirus	Jansons and Bucens 1986
Tap	100	Hollow fiber (10,000)	>2.5	100–750 mL	148% poliovirus	Bicknell, Dziewulski, and Sturman 1985
Tap	100	Hollow fiber (50,000)	1	10 mL	63% phage MS2 55% phage PP7 65% poliovirus 57% *E. coli* 57% *B. subtilis* 26% *C. parvum*	This study
Surface	100	Hollow fiber (10,000)	>3.5	100–750 mL	35% poliovirus	Bicknell, Dziewulski, and Sturman 1985
Sea	100	Spiral cartridge (30,000)	~2	130 mL	>80% phytoplankton viruses (PWH3a-P1 and LMG1-P4)	Suttle, Chan, and Cottrell 1991
Tap	50	Hollow fiber (10,000)	3.5	75 mL + backwash	52% poliovirus	Belfort, Rolem, and Katzenelson 1976
Tap	50	Hollow fiber (68,000)	~4	140 mL	69% phage MS2 47% *E. coli* 44% *C. parvum**	Juliano and Sobsey 1998
Deionized tap	20	Polysulfonate cassette	0.3	>150 mL beef extract elution	96% poliovirus 83% echovirus	Garin et al. 1993
Surface	10	Hollow fiber (50,000)	ND	250 mL	59% phage T1 63% phage PP7 92% *E. coli* 32% *C. parvum*†	Morales-Morales et al. 2003
Tap	10	Flat bed (50,000)	ND	ND	82% poliovirus 14% *G. lamblia*	Kfir et al. 1995
Ground	2	Hollow fiber (50,000)	1.5	50 mL	90% poliovirus 88% phage PP7	Winona et al. 2001
Tap	2	Hollow fiber (50,000)	~1.5	<100 mL	65% *C. parvum*	Kuhn and Oshima 2002

*Juliano and Sobsey recoveries occurred after centrifugation step.
†*C. parvum* percent recovery includes IFA–IMS step and is not representative of recovery from UF step alone.
ND = no data.

Bacterial recoveries in the primary UF step were 87.5% and 98.9% for *B. globigii* spores and *E. coli* K-12 vegetative cells, respectively (prior to centrifugation). Clumping of bacterial spores has been observed and may be responsible for a lower average recovery efficiency of *B. globigii* spores as compared to that of *E. coli* vegetative cells in the initial UF. The condition of the spores should be monitored microscopically to determine the condition of the spore preparation prior to use.

For viral agents, the results indicate that the most challenging component of the concentration process was the initial UF step but that viral recoveries were very efficient in subsequent steps. Disaggregation of viral particles during the downstream process may have contributed to the efficient recoveries and the artificially low recoveries in the UF process.

Multiple approaches for reducing microbial association with the filter and particulates and promoting elution of these agents postfiltration appears to be beneficial. The pretreatment of filters with blocking agents, coupled with the introduction of amendments of Tween 80, NaPP, and proteinaceous agents to the initial sample, has improved recoveries. Addition of glycine or other elution agents is useful in promoting the release of bound agents from the filter and particulates in the concentrate. Multiple treatments in a single process may serve to minimize the effect of water quality variation and differences between microbial agents and may result in more consistent recoveries. The addition of NaPP to Tween 80 as an amendment increased the recoveries of several of the microbial agents and also reduced the variation in performance (Table 3.3).

Microbial association with filter surfaces occurs through hydrophobic and electrostatic interactions that can be influenced by several factors. These factors include pH, salinity, particulate matter, and organic compounds (Gerba et al. 1978, Guttman-Bass and Armon 1983, Guttman-Bass and Catalano-Sherman 1985, Shields and Farrah 1983, Shields et al. 1986, Sobsey and Glass 1984, Sobsey and Hickey 1985). Such factors affect microfiltration as well as UF processes. The interaction between microbes and surfaces remains a complex and poorly understood relationship that can vary between water samples. The cumulative effects of multiple treatments will help reduce variation between water samples and microbes.

Stepwise and Cumulative Recoveries

Concentration from 100 L to the final concentrate is a multistep process, and each step contributes to the overall recovery efficiency. In this study, microbe recoveries were obtained at the end of each major process in order to determine the contribution of individual steps to the overall concentration process. Viral agents are particularly vulnerable because of the extended concentration process as compared to cellular pathogens.

The efficiency of the overall viral recoveries was limited by the first UF process because the initial UF tended to have the lowest average stepwise recoveries. Step recoveries showed that when the initial UF was finished, the rest of the process yielded little loss of virus. In fact, some of the processes produced stepwise recoveries >100%. Viruses may have formed aggregates with other viruses and/or interacted with particulates in the sample during the concentration process. If aggregation was occurring, the retentate sample would underrepresent the actual amount of viruses present because the plaque assays could not differentiate between one virus and many aggregated virions. The *g* force applied during the spin-column step may have disaggregated viral clumps in the sample. Similar observations were reported for the concentration of surface water (Olszewski, Winona, and Oshima 2001).

The initial filtration step prior to centrifugation produced much greater average recovery efficiencies for bacterial agents when compared to recoveries from the centrifugation pellet, which was 54.1% and 50.4% for *B. globigii* and *E. coli*, respectively. Interestingly, when supernatant and pellet recoveries were combined, 90.4% of *E. coli* was recovered, whereas only 72.6% of *B. globigii* was accounted for, an indication that some of the cells were remaining in the supernatant. Limited tests showed increased centrifugation speeds or that the use of a second centrifugation step did not increase recoveries for *B. globigii* spores (data not provided).

Microfiltration of the initial retentate may be an effective alternative to the centrifugation step, thereby allowing isolation of cellular organisms without the threat of clumping or cellular damage while removing many of the particulates in the sample prior to secondary UF. The process also eliminates the need for centrifugation and can potentially be a more rapid process than centrifugation. Potential loss of virus and recovery efficiencies of cellular organisms must be assessed before this strategy can be recommended.

For simplification of the overall process for viral concentration, experiments were conducted where the concentrate proceeded immediately to a secondary UF process, thereby skipping the centrifugation step. This resulted in low average stepwise viral recovery of 24% (n = 2) from the secondary retentate and increased the amount of time required for filtration by ~15 minutes. Presumably, the centrifugation step removed particulates that would interfere with the speed and efficacy of the secondary UF step. It is also possible that centrifugation in the presence of elution agents resulted in the release of bound viruses during the process. Improved recoveries after centrifugation have been observed in surface water concentrates (Olszewski, Winona, and Oshima 2001).

The average final cumulative recovery for *C. parvum* was 25.9%. Other research has demonstrated a 44% recovery efficiency of *C. parvum* from 50 L of tap water using a UF approach (Juliano and Sobsey 1998). Using an equivalent filter and pump system, Kuhn and Oshima (2001) achieved 65% recovery of seeded oocysts from 2 L of surface water without using an IMS step. The lower recovery efficiency of *C. parvum* as compared to other organisms used in this study may be due to low recovery from the initial UF step, the centrifugation step, or from the IMS–IFA assay, yet is likely to be combination of the three (Kuhn and Oshima 2002). The purification and enumeration of *C. parvum* oocysts was the most complex process compared to that of other organisms.

To maximize the volume sampled, the entire pellet was used for IMS–IFA purification and detection. The pellet was, in some cases, a much larger volume than the 0.5-mL equivalent of the packed pellet recommended in USEPA Method 1623 (2001b). Previous research demonstrated that protozoan recovery using IMS decreased as turbidity increased (Bukhari et al. 1998). The pellet generated from the centrifugation step contained particulate matter concentrated from 100 L of water. The particulates that were present substantially increased the turbidity of the sample and could have interfered with the efficiency of the IMS step, thereby contributing to low recoveries of *C. parvum* in the enumeration assay. Previous research with 10-L tap water samples and a smaller surface area Microza filter demonstrated an average filtration recovery of 60.9%, average IMS recovery of 62.4%, and average overall recovery of 38.3%. Substantial loss of oocysts in the enumeration assay would foster the underestimation of UF recovery efficiencies of *C. parvum*. However, the reported recoveries for 100 L were not that much lower than that of the 10-L system. Recoveries reported in this study are a conservative estimate of *C. parvum* oocyst recoveries from 100 L because of the use of a larger pellet volume than recommended for the IMS process. However, this approach demonstrated the utility of using the entire pellet from a 100-L sample, which may be desirable for increasing the ability to detect dilute concentrations of biological agents.

Recoveries generated in this study compare favorably to previous studies (Table 3.13). Comparisons, though, are difficult to compare accurately because of differences in target organisms, degrees of concentration, water volumes, and water qualities.

A method was also developed for long-term (<1 month) storage of blocked Microza ultrafilters. When water supplies are suspect and rapid diagnostic methods are required, the addition of 0.01% thimerasol to the blocking solution facilitates the storage of filters for subsequent immediate use (no sanitation or further blocking required). Results of the storage experiments also indicate that filters could be stored without thimerasol when no growth was detected. However, great care must be taken to avoid contamination if this approach is used. Although storage experiments were not performed with 100-L Microza filters or IMT filters, a similar approach should be feasible. Although not tested in this study, it may be feasible to extend the storage times to even longer periods.

This study developed a capable method for reducing 100 L of drinking water to 10–15 mL of concentrate following centrifugation for the detection of cellular organisms and to 15–20 mL of final concentrate for the detection of viral agents. The final concentrate for cellular organisms was produced in 1.5 hours and in 3 hours for viral agents.

Additional concentration processes for viral agents were necessary after the initial UF process. The use of a secondary UF step followed by spin-column filtration proved to be the most effective process. Greater than 88% of all the viral agents present in the concentrate after the initial UF step were recovered in the final concentrates. This process was also the most efficient, producing a final concentrate within 3 hours.

The NanoCeram fibers were used as an alternative to the secondary UF process. The recoveries were efficient; however, the process was more time consuming and required the use of centrifugation. The results indicate that binding of viral agents to the fibers was very efficient even in the presence of Tween 80 and glycine; therefore, this approach is worth considering as a possible component in the concentration of viral agents.

The IMT ultrafilters produced similar results to the Microza ultrafilters for 10-L samples; however, additional study should be conducted on the IMT filters for 100 L of tap water. The higher cost for the Microza filter ($1,500) necessitates its regeneration and reuse, whereas the lower cost for the IMT filter ($275) allows its use as a disposable filter. Another main expense is the peristaltic pump and pump head (~$2,500), which would be necessary for both filtration systems. The remaining significant expense is the cost of the spin columns at ~$24 per sample.

When biological agents are introduced into drinking water sources or facilities, rapid detection is critical for the implementation of safety procedures to ensure public health. The rapidity and potential automation of the preconcentration process fosters its use as a method for the detection of biological agents in drinking water distribution systems.

The results of this study show that UF coupled with downstream concentration methods can provide a viable approach for the rapid, effective, and simultaneous concentration of viruses, bacteria, and protozoa from large volumes of drinking water.

CONCLUSIONS

An UF-based preconcentration process for viral, bacterial, and protozoan agents in water was developed and optimized. Recoveries were greater than 50% for bacteria and viruses, and greater than 25% for *C. parvum* for the complete process. Specific conclusions are as follows:

1. The use of Tween 80 and NaPP as amendments in the initial sample as well as the use of 0.05*M* glycine at pH.7 in the elution process improved recovery efficiency for the Microza 50,000 Da MWCO hollow fiber ultrafilter.
2. The storage of blocked filters for up to 4 weeks is feasible with either sterile DI water or 0.01% thimerasol added to the blocking agent. No bacterial contamination was detected in the filter module during the course of storage with either treatment. Storage of blocked filters is a viable option because it allows the immediate use of the filters once a sample is obtained.
3. The efficient concentration of viral, bacterial, and protozoan agents from 100 L of tap water to 15–20 mL is possible. The most efficient process for the recovery of microbes involved the use of two UF steps and the use of spin columns (phage PP7, 55.2%; MS2, 72.3%; poliovirus, 85.8%; *B. globigii* spores, 54.1%; and *C. parvum*, 25.9%). The entire process can be completed in <3 hours.
4. An alternate downstream concentration method is the use of adsorptive fibers (Nano-Ceram fibers) and spin columns following the initial UF process (Table 3.9). The process, however, is more time consuming than the use of the secondary UF process and spin columns (~1 hour longer).
5. The IMT hollow fiber ultrafilter offers a potentially lower cost alternative to the Microza hollow fiber filter. Efficient recoveries were produced for 10-L volumes with the use of 0.1% NaPP (Table 3.12).

APPENDIX A
10-L FILTRATION PROTOCOL FOR MICROZA ULTRAFILTERS

Objective: The objective of this protocol is to isolate multiple organisms (viruses, bacteria, and protozoa) from a 10-L tap water sample.

Terms:
In crossflow mode, the permeate port is closed.
In filtration mode, the permeate port is open and has a flow rate of 400 mL/min.

NOTE: The storage solution may vary, depending on the experiment. Currently, the storage solution consists of sterile water. If storage solution is 0.01% thimerasol in the blocking solution, follow the steps indicated with an asterisk (*) in appendix C.

SANITIZATION

Sanitization should take place the day before filtration according to the following steps:

1. Obtain filter from 37°C incubator.
2. Remove all caps, empty filter of sodium dodecyl sulfate (SDS), and cover side ports. Attach the filter directly to the reverse osmosis (RO) water faucet with a 4–5-in. piece of hose to the retentate port.
   a. Rinse filter for 20 minutes in crossflow mode to remove 10% SDS, or fill carboy with 140 L of water and rinse using a pump that has been set up with appropriate hoses.
   b. To attach hoses to the filter, attach an input hose that is equipped with a psi gauge to the bottom of the filter (input port). Attach retentate tubing to the top of the filter (retentate port), which is also equipped with a psi gauge.
   c. To attach hoses to the pump, thread the input hose (for sample) through the pump (place hose in clockwise position), clamp the hose to the pump, and tighten the screw. Leave approximately 2–4 in. of tubing between pump and psi gauge, being careful not to tighten the screw excessively.
   d. Place the filter and hoses on a ring stand, and filter the water at speed 5 under crossflow mode.
3. Prepare a sanitizing solution (0.5*M* NaOH (10g NaOH/500 mL water) + approximately 2.5 mL bleach to obtain at least a 200 ppm (2.0 mg/L) chlorine (Cl_2) concentration).
4. Check the free chlorine level in the sodium hydroxide–chlorine ($NaOH/Cl_2$) solution as follows:
   a. Turn on the spectrophotometer (Hach DR/2010) and use program 80. Zero-out the spectrophotometer by placing a blank (10 mL deionized [DI] water) in the first cuvette. Place the blank in the spectrophotometer and press the "zero" button.
   b. Measure 9.9 mL of DI water in a 15-mL centrifuge tube. Add 100 μL of the $NaOH/Cl_2$ solution to the sample, and add a Hach powder pillow (catalog number

21055-69). Mix well, creating a vortex. Measure the sample immediately by placing it in the spectrophotometer and pressing the "read" button.

c. The spectrophotometer reading for the second cuvette should be 2.00 (which is equivalent to 200 ppm). If the solution is less than 2.00, more bleach must be added to the sanitizing solution. If the reading is greater than 2.00, perform steps 12–16 for a longer period of time to remove all the chlorine.

NOTE: The concentrated bleach must be tested after each use to ensure that enough bleach is being used to reach 200 ppm. Rinse cuvettes after each use, being careful not to touch the cuvettes because fingerprints may cause an inaccurate reading.

5. After rinsing the filter, attach the input and retentate hoses to the filter and the pump (if the DI water faucet was used), as described in steps 2b–d and place equipment in the biological safety cabinet. If the pump was used, place the pump equipment in the safety cabinet and proceed to step 6.
6. Place the input and retentate hoses into the sanitizing solution.
7. Circulate the sanitizing solution in crossflow mode through the filter for 1 hour at speed 5.
8. After circulating the sanitizing solution, empty the sanitizing solution from the filter by lifting the input and retentate tubes above the remaining solution while the pump is still running (the solution will exit from the retentate hose). Repeat this step by placing the pump in the reverse position (the solution will exit from the input hose).
9. Remove hoses from the pump and put pump away.
10. Discard any used sanitizing solution appropriately.
11. Attach the input hose to the RO water faucet, and rinse the filter in filtration mode but without using the pump.
12. Place both permeate and retentate hoses in the sink. Place a pinch clamp for backpressure on the retentate tube ~1 to 1½ in. above the pressure gauge.
13. Turn on the RO faucet (or use the pump) and set the filtration parameters for filtration mode (pump speed 5, psi ≤15, flow rate 400 mL/min).
14. Use a graduated cylinder to measure the permeate flow for 1 minute. Allow the permeate to stabilize, and begin measuring/timing after 100 mL has been collected. At the end of 1 minute, record the flow rate.
15. Allow the filter to be rinsed until the chlorine level is below 0.01 mg/Cl_2, which will take ~30–60 minutes and may require up to 480 L of water. Check the free chlorine by repeating step 4, but this time, obtain the sample from the permeate port.

    NOTE: The reading on the spectrophotometer for the second cuvette should be 0.01, which is equivalent to 1 ppm. If the concentration is more than 1 ppm, continue rinsing with more water.
16. After rinsing the filter, check the pH of the permeate port to ensure that it ranges from 7.0 to 8.0 (typical house RO water is pH 8).
17. Proceed to the Blocking section of this protocol.

BLOCKING

After sanitization, the filter must be blocked. Blocking is to take place the day before filtration according to the following steps:

1. Remove all excess water from the filter by removing all caps and shaking gently.
2. Prepare a 5% solution of calf serum with thimerasol (475 mL DI water + 25 mL calf serum + 0.5 g thimerasol). *Gently* mix the stock calf serum prior to use.
3. Set up the pump in crossflow mode according to steps 2b–d of the sanitization protocol. Place the input and retentate hoses in the blocking solution.
4. Allow this solution to circulate under crossflow mode through the filter for 1 hour with the pump at speed 5 (psi <15).
5. After circulation, remove the filter and hoses from the pump.
6. Re-cap all ports to the filter and fill the pump with approximately 25–50 mL of the blocking solution just used (or as much will go into filter).
7. Place the filter on a shaker overnight in a cold room (4°C) for a minimum of 12 hours but no more than 24 hours.
8. Proceed to the Filtration section of this protocol.

FILTRATION

Filtration is to take place the day of filtration according to the following steps:

1. Obtain a 10-L sample and place the necessary materials in the safety cabinet (filter, pump, tubing, clamps, water sample, etc.).
2. Perform the following steps for preparation of the 10-L sample:
   Add 1,100 mL sterile 10× phosphate-buffered saline (PBS) to the sample and 1 mL Tween 80. Collect a 10-mL sample (subsample). Then add any test organisms (i.e., 270 μL 10^6 PP7) at their appropriate concentration (see appendix G for *Escherichia coli* protocol). Mix well with the lid on the carboy. Collect a 10-mL sample of this mixture prior to filtration (input sample).
3. Remove the filter from the cold room. Collect a 10-mL sample of the blocking solution to test for any bacteria that are in the filter prior to filtration. Rinse the filter in crossflow mode with 10 L of DI water to rinse out any residual thimerasol.
4. Attach the hoses to the filter and then to the pump inside the safety cabinet according to steps 2b–d in the Sanitization section.
5. Place the permeate tubing in an empty, sterile carboy, and place the retentate and input hoses in the carboy with the water sample. Tape the hoses to the carboy to prevent them from sliding out.
6. When the water begins to flow through the retentate port, begin setting the parameters for the filtration mode.
7. Use a graduated cylinder to measure the permeate flow for 1 minute. Allow the permeate to stabilize and begin measuring/timing after 100 mL has been collected. At the end of 1 minute, record the flow rate. Adjust the backpressure clamp 1 to 1½ in. from the top of the gauge on the retentate tubing to obtain a 400-mL flow rate. Record the pressure, which should not exceed 15 psi.

8. When the parameters have been adjusted to the desired settings, note the time (or use a timer on "count up") and *do not* try to readjust the flow rate once it has been established.
9. Allow the sample to filter until 200 mL remains (this should take ~30 minutes) and *note the time*. To do this, lift the input and retentate tubes above the remaining liquid and turn the pump on again to empty it (water will exit from the retentate hose). Repeat this step by placing the pump in the reverse position (water will exit from the input). Mix the remaining sample well and pour the ~200–500 mL of concentrate from the carboy into a 500-mL bottle and continue with filtration until ~75 mL remains in the bottle. Lift the input and retentate tubes above the remaining liquid and turn the pump on again to empty it (water will exit from the retentate hose). Repeat this step by placing the pump in the reverse position (water will exit from the input hose).
10. Record the time.
11. Remove the backpressure clamp and the permeate hose. Place a cap on the permeate port. Collect a 10-mL sample of the permeate (permeate sample). The remaining permeate can be disposed of in the sink.
12. Add 20 mL 0.5*M* glycine to the ~200 mL of retentate for a final concentration of 0.05*M* (1:10 dilution).
13. Allow the glycine/sample mixture to circulate with the input and retentate hoses for 30 minutes under crossflow mode at speed 5. Place the bottle with the glycine/sample mixture into a larger container to prevent it from tipping.
14. After 30 minutes, remove all liquid (~220 mL) from the filter by using the forward and reverse modes of the pump (according to step 9 in this section).
15. Mix the concentrated sample well prior to further analysis. Collect a 10-mL sample (final concentrate sample).
16. If centrifuging the sample, proceed to steps 17–20. If not, proceed to the After Filtration section.
17. If centrifuging the sample, place ~220 mL of concentrate in two 175-mL conical tubes and weigh the tubes to ensure they are of equal weight. Centrifuge the sample at 1,240× *g* for 20 minutes.
18. After centrifugation, carefully remove the lids of the centrifuge tubes so that the pellet is not disturbed. Aspirate the supernatant and place it in a bottle for viral analysis. Collect a 10- mL sample of the supernatant.
19. The pellet (for protozoa analysis) must be resuspended in DI water for a final concentration of 0.1 mL per packed pellet per milliliter. For example, if there is a 1-mL packed pellet, resuspend it in enough water to bring it to a final volume of 10 mL. Place the resuspended pellet in the refrigerator until it can be used for immunomagnetic separation.
20. Proceed to the After Filtration section of this protocol.

AFTER FILTRATION

The following steps are to take place after filtration on the day of filtration:

1. After removing glycine from the filter, repeat steps 3–16 in the Sanitization section.
2. Make note of the flow rate compared to the pressure achieved in the Filtration section of this protocol. Also, check previous flow rates in comparison to pressure.
3. If a high pressure (approximately >10 psi) is needed to obtain the 400 mL/min flow rate, proceed to the Acid Wash section of this protocol.

ACID WASH

An acid wash is performed after every filtration on the day of filtration.

1. After sanitizing and rinsing the filter, circulate the filter under crossflow mode with 1% hydrochloric acid (HCl) solution (495 mL DI water and 5 mL 12*M* HCl) for 1 hour with the pump according to step 4 in the Sanitization section (except that acid will be used instead of NaOH/Cl_2).
2. After 1 hour, remove the acid solution from the pump using the forward and reverse modes of pump.
3. Remove the filter and hoses from the pump and attach the input hose to the DI water faucet. Place the retentate and permeate hoses in the sink and rinse filter under filtration mode and the parameters in step 4.
4. When water begins to flow through the retentate port, begin setting the parameters for the filtration mode with a permeate flow of ~300 mL/min after the pinch clamp has been adjusted).
5. After ~30–60 minutes, check the pH of the permeate flow and rinse until it ranges from 7.0 to 8.0.
6. When the acid wash is done, perform another clean water flux according to step 2 in the After Filtration section.
7. If it meets specifications, fill with 50 mL of 10% SDS and store in a 37°C incubator for a minimum of 12 hours until next use.

Table A.1
Necessary reagents and materials

	Supplier	Catalog number
Reagents		
10× PBS pH 7	Sigma, St. Louis, Mo.	P-8416
• Potassium phosphate (KH_2PO_4)	VWR, West Chester, Pa.	4058-07
• Sodium chloride (NaCl)	Sigma	S-5136
• Sodium phosphate (Na_2HPO_4)		
Sodium hydroxide (NaOH)	Fisher Scientific, Pittsburgh, Pa.	BP359-212
N,N-diethyl-p-phenylenediamine free chlorine powder pillows	Hach Permachem Reagents, Loveland, Colo.	21055-69
Calf serum	Gibco, Grand Island, N.Y.	12485-025
Glycine pH 7	Fisher Scientific	BP381-1
Ultrapure SDS	ICN Biomedicals, Cleveland, Ohio	811030
Materials		
1.0-µm PTFE filter membrane, 47 mm	Cole-Parmer, Vernon Hills, Ill.	SR-02916-74
0.2-µm PTFE filter membrane, 47 mm	Cole-Parmer	SR-02916-54
Permeate tubing	Cole-Parmer	96440-36
Retentate tubing	Cole-Parmer	96440-36
Input tubing	Cole-Parmer	96440-36
10-L Microza filter	Pall Corp., Port Washington, N.Y.	77250-62

*PTFE = polytetrafluoroethylene.

APPENDIX B
PROTOCOL FOR INOCULATING AND GROWING BACTERIA (*ESCHERICHIA COLI* K-12) (REVISED NOV. 9, 2003)

Supplies Needed:

- Nutrient broth in 3-mL quantities
- Nutrient agar plates (or trypticase soy agar [TSA] plates)
- 2-mL Eppendorf tubes (Fisher Scientific, Pittsburgh, Pa.)
- Sterile saline
- Autoclave filter units and syringes for membrane filter counts

Stock Inoculum Preparation:
All bacterial transfers should be performed in the biological safety cabinet. Streak the culture on appropriate agar medium to obtain well-isolated colonies and then perform the following steps:

1. Ensure that the plate has uniform colony morphology, and pick an isolated colony to inoculate duplicate 3-mL tubes of nutrient broth.
2. Loosen the caps and incubate at 37°C in an orbital shaker at 150 rpm. After 17 hours of growth, the tubes should have turbid cultures. Date and store the cultures at 4°C. (A new stock inoculum will be ready after 1 month of running stock use.)

Inoculation:
Inoculation is to be performed as follows, before the bacterium is needed:

1. Before each use, the stock culture must be struck onto a plate and examined for purity.
2. Obtain two 3-mL nutrient broth tubes and inoculate with 75 μL of liquid stock inoculum.
3. Loosen the caps and incubate the tubes at 37° for 17 hours. (One tube is to be used in the experiment, and the other is to be stored to serve as a running stock.)

NOTE: If a different volume of bacterium is needed, adjust the amount of inoculum to a 2.5% ratio of inoculum to nutrient broth.

DAY OF FILTRATION

Before Filtration

1. With absorbance set at 620 nm, allow the spectrophotometer to warm up for 15 minutes, then read the absorbance of each tube, using a blank with only the nutrient broth. Whirl to create a vortex in each tube before recording its absorbance. Record the optical density (OD). Typically, the OD should read between 0.5 and 0.7.
2. Using Eppendorf tubes, perform serial dilutions for one of the tubes just incubated. Use nutrient broth for dilutant. The absorbance reading should be similar to those on the growth curve (absorbance = 0.5–0.6). With this absorbance, typically, there are 10^8

to 10^9 per mL, and 0.1 mL of the 10^{-2} dilution can be added to the carboy. This amounts to a total of 10^6 bacteria that are being added for the input.

3. Continue to culture the 10^{-5} dilutions to the 10^{-7} dilutions using 0.1 mL on duplicate plates. Discard the dilution tubes after use. These plate cultures are to be counted the following day.

After Filtration

Analyze each of five samples (blocking solution, input, subsample, permeate, and final concentrate) for bacterial and *E. coli* growth according to the following procedure:

1. Culture 0.1 mL of each of these samples onto duplicate TSA plates. The final concentrate must be serially diluted with saline to ~10^{-3} to obtain appropriate counts. For input samples, proceed to the special enumeration method described in step 3.
2. Because spreading 0.1 mL doesn't give enough cells to be statistically representative when measuring input samples of 10^5/L through 10^2/mL, the membrane filtration technique must be used by attaching the syringe barrel to the filter apparatus and adding 10 mL of sterile saline, followed by 1 mL of input sample. Mix this solution thoroughly by repeatedly depressing the pipette. Next, insert the plunger to force all liquid through the filter membrane.
3. Remove the filter from the filter holder and place it on a TSA plate. For input concentrations of <10^5/L, collect a 10-mL sample using the same technique described in step 1 of this section.
4. Incubate all plates overnight in a 37°C incubator.

DAY AFTER FILTRATION

Calculations

To calculate the input (using dilutions of inoculum):

$$\frac{\text{average number of colonies on plate} \times \text{dilution factor}}{\text{volume on the plate}} = \text{cfu/mL}$$

To calculate the input and final concentrate:

$$\left(\frac{\text{average number of colonies on plate} \times \text{dilution factor}}{\text{volume on the plate}}\right) \times \text{amount*}$$

$$= \text{total number of colonies}$$

* For example, the input is 10,000 mL.

Note any growth on the storage solution plates.

APPENDIX C
10-L FILTER STORAGE PROTOCOL

Objective: Long-term storage (up to 4 weeks) of blocked 10-L filters

NOTE: The storage solution may vary, depending on the experiment. Currently, the storage solution consists of sterile water. If storage solution is 0.01% thimerasol in the blocking solution, follow the steps indicated with an asterisk (*).

DAY BEFORE STORAGE

1. Follow the standard 10-L filtration protocol through the overnight blocking step (in appendix A).
   *Add 0.01% thimerasol into the blocking agent. Place on shaker overnight, omitting steps 2 and 3 of this procedure as well as all of "Day of Storage" steps, and rinse with 10 L of sterile deionized (DI) water before filtration on the day of filtration.
2. Sterilize the 4-in. (6-cm) tubing used to cap the ends of the filter as well as any tubing that will be used for rinsing the filter the following day.
3. Sterilize one 10-L carboy of water for each filter being stored. Sterilize ~200 mL of water that will be used as a storage solution for the filters.

DAY OF STORAGE

All methods must be performed in the biological safety cabinet.

1. Uncap the input and retentate ports and collect a 10-mL sample of storage solution. Plate-out 0.1 mL of this solution onto duplicate trypticase soy agar (TSA) plates in order to monitor any bacterial growth.
2. Attach sterile input and retentate tubing, complete with psi gauges, to each port, respectively. Thread the input tube through the pump (place the hose in a clockwise position to clamp the hose onto the pump, then tighten the screw). Allow 2–4 in. (4–6 cm) of tubing between the pump and psi gauge.
3. Place the input hose in a 10-L carboy filled with sterile water. Place the retentate hose in an empty carboy.
4. Turn pump on, and filter 10 L of the sterile water under crossflow mode (permeate port closed) until all water has passed through the pump.
5. Detach the pump from the tubing. Remove the retentate tubing from the filter and replace it with a 4-in. (6-cm)-long piece of sterile tubing. Secure the opening with a metal clamp.
6. Turn filter over and remove input tubing from the filter.
7. Fill the filter with ~50 mL of fresh, sterile water. Attach a sterile 4-in. (6-cm) piece of tubing to the input port, clamping it shut.
8. Store the filter in a cold room (4°C).

Periodic Monitoring of Storage Solution

The storage solution must be tested in the safety cabinet twice a week according to the following procedure:

1. Sterilize the filter caps, including the extra, clearly marked (injection) caps that will be used to test the storage solution. NOTE: The injection caps are typical filter caps used only to test the storage solution and the only caps punctured with a needle.
2. Obtain a filter from the cold room, and shake the storage solution in the filter.
3. In the safety cabinet, remove either the input or retentate port's 4-in. (6-cm) tubing and replace it with an injection cap.
4. Turn filter over and secure the filter on a ring stand.
5. Attach a sterile 1-cc syringe to a sterile disposable 26½-gauge needle.
6. Insert the needle through the injection cap. Pull back on the syringe to remove ~0.35 cc of storage solution from the filter. Remove the syringe from the injection cap.
7. Placing the needle in a vertical position with the needle at the top, tap the syringe to dislodge the air bubbles, then push fluid through the syringe until all air is removed and fluid remains at the 0.2-cc mark.
8. Invert the syringe and deposit 0.1 cc of storage solution onto duplicate TSA plates. Label appropriately and place into the 37°C incubator.
9. Turn filter over, remove the injection cap, and replace it with a sterile filter cap.
10. Turn filter over again, remove 4-in. (6-cm) tubing, and replace it with a sterile filter cap.
11. Store the filter in a cold (4°C) room until time for the next sampling.
12. Repeat steps 1–11 each time the filter is to be tested for bacterial growth. Each time a sterile injection cap is necessary, replace it with the original filter cap.

Filtration

1. Run a standard filtration procedure and rinse with 10 L of sterile DI water before filtration begins (filters have been stored with 5% calf serum and 0.01% thimerasol). For filters stored in DI water, after the blocking step, rinse with 10 L DI water, drain the water and add 50 mL of sterile DI water, and store in a cold (4°C) room. (NOTE: Filters stored in DI water were ready for immediate use without additional rinsing.)
2. Two or four weeks after storage at 4°C, the filter is removed and filtration is performed by the standard procedure described in appendix A.

APPENDIX D
OVERNIGHT *BACILLUS GLOBIGII* SPORE PURIFICATION (REVISED MAY 25, 2004)

Materials Needed:

- Generic spore media (GSM)—for 1 L
  - Nutrient broth: 8 g
  - Manganese sulfate: 120 mg
  - Calcium chloride ($CaCl_2$): 300 mg
  - Deionized water: 1,000 mL
  - Tubes (3 mL) or flasks (30 mL)
- Twenty 15-mm screw-cap tubes
- Two 40-mL Oakridge tubes
- Spectrophotometer
- Two 125-mL sidearm flasks
- Sterile loop
- Bunson burner
- 100-µL pipette
- Centrifuge
- Water bath
- Cold, sterile H_2O
- 40% ethyl alcohol
- 10 m*M* Tris with lysozyme:
  - 0.25 mL of 1*M* Tris in 24.25 mL of sterile water (i.e., 25 mL of 10 m*M* Tris)
  - 25 mg of lysozyme
- 0.1% Tween/saline solution
- Fresh plate of *B. globigii* colonies

DAY 1

All of the following steps must be performed in a biological safety cabinet using sterile technique.

1. With a pipette, add 30 mL of GSM to each of two 125-mL sidearm flasks.
2. With a sterile loop, pluck a colony from the *B. globigii* plate.
3. Inoculate each flask of 30 mL GSM with the colony, tighten the lids of the sidearm flasks, and gently swirl each flask.
4. Loosen the lids of the sidearm flasks.
5. Place the flasks in a 37°C orbital shaker.
6. Incubate the cultures for 20 hours.

DAY 2

1. Turn on the spectrophotometer 15 minutes prior to use. Heat water bath to 70°C.
2. Prepare a 15-mm screw-cap tube containing 3 mL of the GSM.
3. After 20 hours of incubation, remove the cultures from the incubator.
4. Take a spectrophotometer reading [at 620], following directions written on the spectrophotometer.
5. Streak a plate and a slide for each culture.
6. Place the plates in the 37°C incubator overnight.
7. Pass the slides through a flame to fix the cells, and set aside for staining.
8. Divide each culture into two 15-mL conical tubes.
9. Place the conical tubes in the 70°C water bath for 20 minutes in order to pasteurize the cells.
10. After pasteurization, place the sample in two sterile 40-mL Oakridge tubes.
11. Centrifuge the samples for 10 minutes at 10,000 rpm.
12. After centrifuging, carefully remove the tubes from the centrifuge without disturbing the pellets.
13. Pipette the supernatant and discard.
14. Add 10 mL of the Tris–lysozyme solution to each tube and resuspend the pellets.
15. Place the tubes in the 37°C incubator for 1 hour. Whirl each tube to create a vortex every 15 minutes.
16. After 1 hour, centrifuge the samples for 20 minutes at 10,000 rpm.
17. Remove the supernatant and discard.
18. Add 10 mL of cold, sterile water to each tube and create a vortex to mix.
19. Centrifuge the tubes at 10,000 rpm for 10 minutes.
20. Carefully remove the supernatant and discard.
21. Add another 10 mL of cold, sterile water and create a vortex to mix.
22. Centrifuge the tubes at 10,000 rpm for 20 minutes.
23. Carefully remove the supernatant and discard.
24. Resuspend the samples in 0.1% Tween–saline solution.
25. Store in refrigerator until ready for use.

APPENDIX E
ADSORBENT FIBER SPIKED CHALLENGE

Objective: Concentration of viral particles into a smaller volume by adsorption and elution of viral particles to a positively charged powder

Materials: Refer to supply list (appendix H) for catalog numbers and vendors.

- *Pseudomonas aeruginosa* slant
- Nutrient broth and soft agar (prepare as directed on label)
- Two sterile 50-mL Falcon (Fisher Scientific, Pittsburgh, Pa.) tubes
- Six sterile 15-mL Falcon tubes
- Disposable sterile loops
- Twenty-seven sterile, soft agar tubes (with 12-mL Falcon polystyrene snap caps)
- Nine sterile dilution tubes (with 5-mL polystyrene snap caps)
- Twenty-seven trypticase soy agar (TSA) plates
- NanoCeram (Argonide Corp., Sanford, Fla.) fibers
- Weigh boats
- Shaker
- Centrifuge
- 5-mL and 10-mL individually wrapped sterile pipettes
- Dechlorinated tap water (50 mg/L sodium thiosulfate) (stock can be kept for ~2 weeks at 4°C.)

DAY 1

Prepare an overnight culture of *P. aeruginosa*, according to the following procedure. (These steps should be performed in a biological safety cabinet).

1. Pipette 35 mL of nutrient broth into a 50-mL conical tube.
2. Retrieve the *P. aeruginosa* slant from the refrigerator.
3. Using a sterile loop, remove a sample from the slant (use the slant no more than three times).
4. Inoculate the nutrient broth with the *P. aeruginosa* contained on the loop.
5. Mix the culture by whirling to create a vortex.
6. Loosen the cap on the 50-mL conical tube.
7. Place the tube in a 37°C incubator overnight.

DAY 2

Prepare a 4-hour culture (in safety cabinet) as follows:

1. Remove the overnight culture from the incubator, and observe its appearance. (The culture should appear turbid with a greenish tint.)
2. Pipette 35 mL of nutrient broth into a sterile 50-mL conical tube.

3. Mix the overnight culture by whirling to create a vortex, and remove 1 mL of culture and add it to the nutrient broth.
4. Mix the 4-hour culture by whirling to create a vortex.
5. Dispose of the overnight culture.
6. Loosen the cap on the conical tube and place it in a 37°C incubator for 4 hours.

Prepare materials for plating (in biological safety cabinet) according to the following procedure:

1. Heat water bath to 45°C.
2. Obtain 27 TSA plates and label them as follows:
   - Input sample plates (two plates for each dilution): I-1, I-2
   - Supernatant 1 plates (two plates for each dilution): S1-0, S1-1
   - Supernatant 2 plates (two plates for each dilution): S2-0, S2-1, S2-2
   - Supernatant 3 plates (two plates for each dilution): S3-0, S3-1, S3-2
   - Resuspended pellet (two plates for each dilution): P-0, P-1, P-2
   - Negative control (one plate): NC
3. Add 2.7 mL of nutrient broth to nine dilution tubes. Place tubes in the refrigerator until ready to plate the samples.
4. Obtain soft agar from refrigerator. Slightly unscrew the cap and place the agar in the microwave oven until the agar is boiling and completely melted.
5. Obtain 27 tubes and add 4.5 mL of soft agar to each tube. Place in water bath immediately so that the soft agar does not solidify.

Prepare a spiked challenge as follows:

1. Add 13 mL of stock dechlorinated tap water to a 15-mL conical tube.
2. Spike the tap water with 5,000 pfu of [10^6] bacteriophage PP7 (270 µL).
3. Mix the sample by whirling to create a vortex for about 15 seconds.
4. Place an input sample (3 mL) in the refrigerator until plaque assay. (NOTE: Samples must be plated immediately after the experiment).

Adsorption:

1. Add 0.5 g of NanoCeram powder to the spiked tap water.
2. Mix the solution by whirling to create a vortex for 15 seconds in order to disperse the powder.
3. Place the tube on the shaker (in a cold [4°C] room) for 15 minutes in order to allow adsorption of the virus to the powder fibers.
4. After 15 minutes, remove tube from the shaker.
5. Place tubes in a centrifuge (in a cold [4°C] room) and centrifuge at 2,500 rpm for 15 minutes. (If centrifugation of a single sample is necessary, place into the rotor an additional 15-mL conical tube, with a mixture of the same weight (±0.05), opposite the sample so the centrifuge is not off-balance.)
6. When centrifugation is complete, remove the tube, being careful not to disturb the pellet. (NOTE: One layer on top of the pellet is acceptable.)
7. With a pipette, carefully transfer the supernatant (S1) to a clean 15-mL (labeled) conical tube and place it in the refrigerator until plaque assay.

Elution:

1. Decide which eluent (10 mL) is to be used and adjust to the desired pH using 1*M* NaOH (sodium hydroxide) or 1*M* HCl (hydrochloric acid) (pH should be above 7.0).
2. Pour eluent into tube containing the pellet with bound virus.
3. Resuspend the pellet in the eluent by vigorously shaking the tube until there is no pellet remaining at the bottom of the tube and the powder is dispersed throughout the eluent.
4. Place the tube on the shaker (in cold [4°C] room) for 15 minutes.
5. Centrifuge the mixture at 2,500 rpm for 15 minutes at 4°C (balance tubes if necessary).
6. When centrifugation is complete, remove the tube, being careful not to disturb the pellet.
7. With a pipette, carefully transfer the supernatant (S2), which should contain the separated virus, to a clean, labeled 15-mL conical tube and place in the refrigerator until plaque assay.
8. If there is not a second dilution step, resuspend the remaining pellet in 10 mL of deionized water and place in refrigerator until plaque assay.
9. If an additional elution is to be performed:
   - Resuspend the pellet in the second eluent.
   - Repeat elution procedure.
   - With a pipette, carefully transfer the supernatant (S3), which should contain the remaining virus, to a clean, labeled 15-mL conical tube and place in refrigerator until plaque assay.

Prepare a bacteriophage plaque assay (in safety cabinet) according to the following procedure:

1. Because the virus must be diluted 1:10 in the previously prepared dilution tubes, dilute 0.3 mL of stock virus into a dilution tube labeled "–1." Whirl to create a vortex. Then pipette 0.3 mL of that initial dilution (–1) into the next dilution tube (–2). Whirl to create a vortex. Repeat this process for all samples.
2. For each dilution, add the following to the soft agar tube: 1 mL of the virus suspension and 0.5 mL of the 4-hour culture (mix by whirling to create a vortex)
3. Pour the mixture onto the TSA plate.
4. Lightly swirl the mixture to evenly distribute the mixture across the plate.
5. Allow plates to harden.
6. Place plates upside down in a 37°C incubator overnight.

DAY 3

Remove plates from the incubator, count the bacterial plaques, determine the virus concentration, and record the results on a virus titer sheet.

- For calculations, use the most accurate plaque counts, which would be between 20 and 200 plaques.
- Plates with >200 plaques are too numerous to count.
- Plates with <20 plaques will increase the error in the calculation and therefore should be avoided.

Calculations

total plaque-forming units in each sample = (average of two plates)(dilution factor)(sample volume)

$$\%\ \text{virus recovery} = \frac{\text{total plaque-forming units in eluent(s)}}{\text{total plaque-forming units in input}} \times 100$$

APPENDIX F
90-MM-DISC SPIKED CHALLENGES

Objective: To concentrate viruses and other organisms from a 10-L tap water sample by a positively charged disc

Materials: Refer to supply list (appendix H) for catalog numbers and vendors.

- Millipore (Billerica, Mass.) stainless-steel filter holder, 90 mm
- NanoCeram (Argonide Corp., Sanford, Fla.) 90-mm filter discs
- Masterflex Tygon (Cole-Parmer, Vernon Hills, Ill.) tubing
- Two sterile 10-L Nalgene (VWR, West Chester, Pa.) carboys
- Masterflex L/S standard drive and pump head
- Nineteen Trypticase soy agar (TSA) plates
- Nutrient broth and soft agar (prepare as directed on label)
- Organism being spiked
- *Pseudomonas aeruginosa* slant
- Three sterile 15-mL Falcon tubes
- Two sterile 50-mL Falcon tubes
- Nineteen sterile soft agar tubes (with Falcon 12-mL polystyrene snap caps)
- Seven sterile dilution tubes (with 5-mL polystyrene snap caps)
- Disposable sterile loops
- Centrifuge
- Sodium thiosulfate

PROCEDURE

Day 1 (Day Before Filtration)

If spiking with bacteriophage PP7, make an overnight culture of *P. aeruginosa*. Perform the following steps in the biological safety cabinet:

1. Pipette 35 mL of nutrient broth into a 50-mL conical tube.
2. Retrieve the *P. aeruginosa* slant from the refrigerator.
3. Using a sterile loop, remove a sample from the slant (the slant should be used no more than three times).
4. Inoculate the nutrient broth with the *P. aeruginosa* contained on the loop.
5. Mix the culture by whirling to create a vortex.
6. Loosen the cap on the 50-mL conical tube.
7. Place the tube in a 37ºC incubator overnight.

The stainless-steel filter apparatus must be autoclaved as follows:

1. Remove legs of apparatus with Allen (Allen Manufacturing Co., Hartford, Conn.) wrench.
2. Before autoclaving, wash the filter unit with soap, rinse with tap water and then reverse osmosis water. Allow to dry.
3. Assemble apparatus according to manufacturer's instructions. Place the 90-mm disc smooth side down on top of the support screen.
4. Seal all openings by wrapping with aluminum foil.
5. Remove the pressure gauge and seal all openings with aluminum foil.
6. Autoclave the apparatus and tubing for 15 minutes on wrapped cycle.
7. After autoclaving, keep all openings sealed until use.

Day 2 (Day of Filtration)

Prepare the 4-hour *P. aeruginosa* culture (in safety cabinet) according to the following procedure:

1. Remove overnight culture from the incubator and observe its appearance. (It should appear turbid with a greenish tint.)
2. Pipette 35 mL of nutrient broth into a new 50-mL conical tube.
3. Mix the overnight culture by whirling to create a vortex, and remove 1 mL of culture and add it to the nutrient broth.
4. Mix the 4-hour culture by whirling to create a vortex.
5. Dispose of overnight culture.
6. Loosen the cap on the conical tube and place the tube in a 37ºC incubator for 4 hours.

Prepare materials for plating (in the safety cabinet) according to the following procedure:

1. Heat water bath to 45°C.
2. Obtain 19 TSA plates.
3. Label the TSA plates as follows:
   - Input sample plates (two plates for each dilution): I-1, I-2, I-3
   - Output sample plates (two plates for each dilution): OP-0, OP-1
   - Backflush sample (two plates for each dilution): BF-0, BF-1, BF-2, BF-3
   - Negative control (one plate): NC
4. In seven of the dilution tubes, add 2.7 mL of nutrient broth. Place tubes in the refrigerator until ready to plate the samples.
5. Obtain soft agar from refrigerator. Slightly unscrew the cap and place agar in the microwave oven until the agar is boiling and completely melted.
6. Obtain 19 tubes and add 4.5 mL of soft agar to each tube. Place immediately in water bath so that soft agar does not solidify.

Preparation of Water Sample:

1. In a 10-L Nalgene carboy, add 10 L of tap water.
2. Dechlorinate the tap water by adding 50 mg/L of sodium thiosulfate. Shake carboy to distribute dissolved sodium thiosulfate throughout water.

3. Spike the water with the desired organism. All organisms used in these experiments should be diluted to the same concentration.
4. Shake the carboy to distribute reagents and organisms.
5. Place a 10-mL input sample into a sterile, labeled 15-mL conical tube. Place in refrigerator until ready to plate (samples should be plated immediately after experiment).

Filtration:

1. Place filter holder in the biological safety cabinet.
2. Ensure that screw caps are tight with the use of an Allen wrench.
3. Attach input tubing with pressure gauge to the inlet port, which is located on the top of the apparatus. Secure tubing with a high-pressure clamp.
4. Open the pump head and wrap the other end of the input tubing over and then under the rotor.
5. The free end of the input tubing should be placed inside the carboy containing the spiked tap water.
6. Attach the output tubing to the outlet port on the bottom of the apparatus. Secure tubing with a high-pressure clamp.
7. Place the free end of the output tubing into a sterile, empty 10-L carboy.
8. Attach a third piece of tubing to the port coming out of the vent/relief valve and place the free end in the sink.
9. Before starting pump, set the pump speed to 0.
10. Start a timer and record how long the filtration takes.
11. Slowly start the pump.
12. Open the vent/relief valve (up position) and watch for water to flow through tubing on vent/relief port. If necessary, slightly tilt the apparatus to ensure that water flows evenly over the disc.
13. Begin increasing the pump speed. (It should not exceed 2 and the pressure should not exceed 5 psi.)
14. When no bubbles are visible in the vent/relief tubing, close the valve (down position).
15. Check the flow rate by timing the output for 1 minute as it flows into a graduated cylinder. Record the volume (in milliliters per minute). The flow rate should be 200–300 mL/min.
16. Continue the filtration process until all the sample has passed through.
    NOTE: Watch for leakage around the filter. If leaking occurs, tighten the screw caps.
17. When filtration is complete, place a 10-mL sample of the output (the water that is now in the carboy) into a sterile, labeled 15-mL conical tube. Place in refrigerator until ready to plate.

Backflush (Elution):

1. Prepare eluent (100 mL) to be used for backflush. Adjust to desired pH with 1*M* HCl or 1*M* NaOH (pH should be above 7.0, in the range of 9.3).
2. Remove the input tube from the pump head.
3. During backflush, the input flows through the bottom of the apparatus and the output flows from the top.

4. Attach the tubing that exits the bottom port (now the input) to the pump head and place the free end into the beaker containing the eluent.
5. Place the free end of the tubing that exits the top port (now the output) into a clean, empty beaker.
6. Turn pump on and allow the eluent to pass through the filter into the clean beaker.
7. Transfer backflush to a sterile 50-mL conical tube and centrifuge at 2,000 rpm for 20 minutes (centrifugation is necessary because some of the filter may be in the backflush)
8. Place a 10-mL sample of the supernatant into a sterile, labeled 15-mL conical tube and place in refrigerator until ready to plate.

Calculations

total plaque-forming units in each sample = (average of two plates)(dilution factor)(sample volume)

$$\%\text{ virus recovery} = \frac{\text{total plaque-forming units in eluent(s)}}{\text{total plaque-forming units in input}} \times 100$$

APPENDIX G
MINIMUM INHIBITORY CONCENTRATION *ESCHERICHIA COLI* PROTOCOL

Supplies Needed:

- Sterile nutrient broth
- Sterile glass test tubes with screw caps
- Thimerasol
- Trypticase soy agar (TSA) plates

Procedure:

1. Prepare a 1% thimerasol solution by adding 0.03 g of thimerasol to 3 mL of nutrient broth. Whirl solution, creating a vortex.
2. Prepare 1:10 dilutions of thimerasol, ensuring that there is at least 10 mL of nutrient broth after dilution (9 mL will be used for the test tubes). Whirl, creating a vortex, after each step.
3. Add 3 mL of the nutrient broth–thimerasol solution to each of three test tubes. Clearly label each tube.
4. Prepare three control test tubes that contain only nutrient broth (no thimerasol).
5. Transfer 75 μL of stock *E. coli* to each test tube. Whirl to create a vortex.
6. Check the purity of the stock being used by removing a loopful of stock and streaking it out for isolation.
7. Loosen the caps to each test tube and place tubes in a 37°C shaker for 17 hours.
8. After 17 hours of growth, record the absorbance at 620 nm.
9. Find the concentration of *E. coli* growth in the tubes by diluting (1:10) the contents of the tube using sterile water and plating out 0.1 mL of each dilution onto duplicate TSA plates.
10. Place plates in a 37°C incubator overnight and enumerate the colonies the next day.

APPENDIX H
SUPPLIES

100-L Filtration:

- Masterflex Drive (Cole-Parmer)
  Model: 7549-32
- Masterflex High-Performance Pump Head I/P (Cole-Parmer)
  Model: 77600-62 (tubing 70, 88, 89)
- Microza Ultrafiltration Module (Flow Solutions)
  Catalog number: AHP-2013
  NOTE: 10-week lead time
- Masterflex I/P89 Tubing (Cole-Parmer)
  Size: 25 ft/pack
  Catalog number: EL-06485-89
- 316L Stainless-Steel Barbed Adapters (Cole-Parmer)
  Clamp Size: 1½ in. × 1½ in.
  Catalog number: BH-30515-12
- Rubber Hose Adapter, 14MPHR-1½"-316L-7 (Tri-Clover)
  Size: 1½ in., 38.1 mm
  Catalog number: 213382
- B2CMP-1½"-316L-7 90-degree Elbows (Tri-Clover)
  Size: 1½ in.
  Quantity: 2
  Catalog number: 9634002937
- B7MP-1½"-316L-7 Tees (Tri-Clover)
  Size: 1½ in.
  Quantity: 2
  Catalog number: 9634003337
- 16AMP-1½"-316L-7 Solid End Caps (Tri-Clover)
  Size: 1½ in.
  Quantity: 1
  Catalog number: 210854
- 13MHHM Hinged Clamp, Heavy Duty With Metal Wing Nut (Tri-Clover)
  Size: 2
  Quantity: 9
  Catalog number: 212196

- Stainless-Steel Sanitary Pressure Gauge (90 mm) (Cole-Parmer)
  Type: Pharmaceutical (0–100 psi)
  Quantity: 2
  Catalog number: A-70002-00
- Support Stands—Tough Rectangular Epoxy (Fisher)
  Base: 6½ × 11 in.
  Rod Length: 36 in.
  Catalog number: 14-670D
- Three-Prong Extension Clamps (dual adjustment) (Cole-Parmer)
  Size: Large
  Catalog number: A-08021-36
- Three-Prong Extension Clamps (dual adjustment) (Cole-Parmer)
  Size: Medium
  Catalog number: A-08021-34
- Hook Connector (Cole-Parmer)
  Catalog number: A-08041-30
- Vacuum Tubing Clamps (Fisher)
  Quantity: 2
  Catalog number: 14-199 [10 pack]
- Nalgene Heavy-Duty Rectangular HDPE Tanks With Covers (Fisher)
  Size: 30 gal (113.6 L), 24 × 18 × 18 in.
  Quantity: 2
  Catalog number: 14-831-330D
- Manual Pressure Valves
  Order From Saunders: http://www.saundersvalves.com
- Jumbosep Filter Inserts (VWR)
  Size: 12/pack
  Catalog number: 29300-800

10-L Filtration:

- Masterflex L/S Standard Drive (Cole-Parmer)
  Model: 07520-40
- Masterflex L/S Pump Head (tubing 15, 24, 35, 36) (Cole-Parmer)
  Model: 77250-62
- Microza Ultrafiltration Module (Flow Solutions)
  Catalog number: AHP-1010
- Masterflex L/S36 Tubing (PharMed) (Cole-Parmer)
  Size: 25 ft/pack
  Catalog number: EL-06485-36

- Masterflex Tygon Tubing (R-3603) (Cole-Parmer)
  Size: 50 ft/pack
  Catalog number: EL-03603-36
- Hosecock Clamps (Regular) (Cole-Parmer)
  Size: Up to ⅔ in. (17 mm) tubing
  Catalog number: A-08028-00
- 2½ in. Industrial Utility Gauges (Cole-Parmer)
  Type: 2½-in. diameter with bottom connection, 0–60 psi, glycerin filled
  Quantity: 1
  Catalog number: A-68046-10
  Or
  2½-in. Industrial Gauges, glycerin filled (Cole-Parmer)
  Size: 0–60 psi
  Catalog number: A-68046-10
- NPT Male Branch Tees, natural PP (Cole-Parmer)
  Size: 10/pack
  Catalog number: A-30610-75
- 316SS, Straight Coupling (Cole-Parmer)
  Size: 1-3⁄16 in. (L), ¾ in. (W), 7⁄16 in. (orifice)
  Catalog number: A-31410-01
- High-Pressure Clamps (Fisher)
  Size: for 5⁄16-in. to ⅞-in. tubing
  Quantity: 2
  Catalog number: 14-198-5B [10 pack]
- Support Stands—Tough Rectangular Epoxy (Fisher)
  Base: 5 × 8 in.
  Rod Length: 20 in.
  Catalog number: 14-670B
- Three-Prong Extension Clamps (dual-adjustment) (Cole-Parmer)
  Size: Small
  Catalog number: A-08021-32
- Nalgene LDPE Wide-Mouth Carboys (Fisher)
  Size: 10 L
  Catalog number: 02-961-65A

Adsorptive Microfilter System:

- Stainless-Steel Filter Holder (Millipore)
  Size: 90 mm
  Catalog number: YY30 090 00

- NanoCeram Discs (Argonide Corporation)
  Size: 90 mm
- Stainless-Steel Hose Connector (Millipore)
  Size: ¼ in. NPTM to ⅜ in. inside diameter
  Catalog number: XX 2504705
- Underdrain Support (Millipore)
  Catalog number: YY2209058
- Three Legs With Caps, 5/64 in. Allen Wrench (Millipore)
  Catalog number: YY2209059
- Vent/Relief Valve, 20 bar (Millipore)
  Catalog number: XX4204708
- O-Ring, Polytetrafluoroethylene (Millipore)
  Size: 4/pack
  Catalog number: YY30 09053
- In-Line Delrin Filter, Holder, 25 mm, Pall Gelman (VWR)
  Size: 6 pack
  Catalog number: 28144-255

Adorptive Powder:

- NanoCeram Fibers [Boehmite (AlOOH)]
  (Argonide Corporation)

Cell Culture:

- Trypsin EDTA (Ethylenediamine Tetraacetic Acid) (Gibco)
  Size: 100 mL
  Catalog number: 25300-054
- 10× Dulbeccos Minimum Essential Medium (MEM) (Gibco)
  Size: 500 mL
  Catalog number: 11430-030
- MEM With HEPES (1×) (Gibco)
  Size: 500 mL
  Catalog number: 12360-038
- MEM Without HEPES (Gibco)
  Size: 500 mL
  Catalog number: 11090-081
- T-150 Flasks (Fisher)
  Size: 50/case
  Catalog number: 10-126-32

- 6-Well Tissue Culture Plates (VWR)
  Size: 50/case
  Catalog number: 62406161
- 24-Well Tissue Culture Plates (VWR)
  Size: 50/case
  Catalog number: 62406-183
- Antibiotic-Antimycotic (Gibco)
  Size: 100 mL
  Catalog number: 15240-062
- HEPES Buffer Solution 1*M* (Gibco)
  Size: 100 mL
  Catalog number: 15630-080
- L-Glutamine (Gibco)
  Size: 100 mL
  Catalog number: 25030-081

Other Equipment:

- Centrifuge (Thermo Electron Corp.)
  Jouan, Type CR3i
  Part number: 11175704
  Chamber diameter: 355 mm
  Max. rpm: 14,000; max. dens: 1.2 g/mL; max. load: 3.7 kg at 4,100 rpm
- Biological Safety Cabinet (Thermo Electron Corp.)
  Type A/B3
  Model: 1186

Reagents:

- Thimerasol (white powder) (Fisher)
  Size: 10 g
  Catalog number: BP2542-10
- Sodium Thiosulfate, pentahydrate, 99.5%, ACS Reagent (Sigma)
  Size: 500 g
  Catalog number: S7143
- Sodium Chloride (NaCl) (VWR)
  Size: 12 kg
  Catalog number: JT40587

- Sodium Phosphate (dibasic anhydrous) (Sigma)
  Size: 500 g
  Catalog number: S-5136
- Potassium Phosphate, monobasic; anhydrous (minimum 99%, plant cell culture tested) (Sigma)
  Size: 1 kg
  Catalog number: P-8416
- Sodium Hydroxide (NaOH) (Fisher)
  Size: 2.5 g
  Catalog number: BP359-212
- Calf Serum (Invitrogen)
  Size: 500 mL (Quantity: 10)
  Catalog number: 16170-078
  Reserve reference number: 618581-52
  Lot number: 349541
- Fetal Bovine Serum (Gibco)
  Size: 500 mL
  Catalog number: 12485025
  Reserve reference number: 61753152
  Lot number: 1153090
- Bovine Serum Albumen (minutes 96%; electrophoresis, pH 5.2; Lyophilized Powder) (Sigma)
  Size: 100 g
  Catalog number: A-3912
- Hydrochloric acid (HCl)
  Size: 500 mL
  Can obtain from biology store room
- Bleach
  Size: 1 gal
  Can obtain from biology store room
- Glycine (Fisher)
  Size: 1 kg
- Sodium Dodecyl Sulfate (SDS) (VWR)
  Size: 1 kg
  Catalog number: IC811030
- Tween-80 (Polyoxyethylene Soritan Monooleate) (Sigma)
  Size: 500 mL
  Catalog number: P-8074

- Formaldehyde 37% by weight in amber glass bottle (Fisher)
  Size: 4 L
  Catalog number: F79-4
- Nutrient Agar (Difco) (Fisher)
  Size: 500 mL
  Catalog number: DF0001-17-0

Miscellaneous:

- Filter Units, PES membrane, sterile, Nalgene, 0.2 μL, 150 mL, 50-mm diameter (VWR)
  Size: 12/case
  Catalog number: 28196-258
- Trypticase Soy Agar (TSA) Plates (Fisher)
  Size: 100/case
  Catalog number: B21283X
- Trypticase Soy Agar (TSA) Slants, BBL (Soybean–Casein Digest Agar)
  Size: 100/case
- Free Chlorine Powder Pillows (Hach)
  Size: 100/pack
  Catalog number: 21055-69
- Redi-Tip Yellow (1 μL–200 μL) (Fisher)
  Size: 1000
  Catalog number: 21-197-8G
- Redi-Tip Blue (1 μL–1,000 μL) (Fisher)
  Size: 1000
  Catalog number: 21-197-8F
- Disposable Sterile Loops (Fisher)
  Size: 250/case
  Catalog number: 130752

SUPPLIERS

Alfa Laval Coach Supply, Fort Worth, Texas
Argonide Corporation, Sanford, Fla.
Cole-Parmer, Vernon Hills, Ill.
Crouch Supply, Forth Worth, Texas
Fisher Scientific, Pittsburgh, Pa.
Flow Solutions, Austin, Texas
Gibco, Grand Island, N.Y.
Hach Chemical Company, Loveland, Colo.
Invitrogen, Carlsbad, Calif.

Millipore, Billerica, Mass.
Saunders, Crane Values, Montgomery, Texas
Sigma, St. Louis, Mo.
Thermo Electron Corporation, Marietta, Ohio
Tri-Clover Coach Supply, Fort Worth, Texas
VWR, West Chester, Pa.

APPENDIX I
CONVERSION FACTORS

Conversion		Procedure	
From	To	Multiply number of	By
grams (g)	ounces (oz), avoirdupois	g	0.03527
kilopascals (kPa)	feet (ft) of hydraulic head	kPa	0.3346
kilopascals (kPa)	pounds per square inch (psi)	kPa	0.14504
liters (L)	gallons (gal)	L	0.2642
liters (L)	ounces (oz), fluid	L	33.81
liters (L)	quarts (qt), fluid	L	1.057
millimeters (mm)	inches (in.)	mm	0.03937
pounds per square inch (psi)	pascals (Pa)	psi	6,895
square centimeters (cm^2)	square inches ($in.^2$)	cm^2	0.155
square meters (m^2)	hectares (ha)	m^2	0.0001
square meters (m^2)	square feet (ft^2)	m^2	10.76
square meters (m^2)	square yards (yd^2)	m^2	1.196

REFERENCES

APHA, AWWA, and WEF (American Public Health Association, American Water Works Association, and Water Environment Federation). 1995. *Standard Methods for the Examination of Water and Wastewater*. 18th ed. Washington, D.C.: APHA.

Belfort, G., Y. Rotem, and E. Katzenelson. 1975. Virus Concentration Using Hollow Fiber Membranes. *Water Research*, 9:79–85.

Belfort, G., Y. Rotem, and E. Katzenelson. 1976. Virus Concentration Using Hollow Fiber Membranes—II. *Water Research*, 10:279–284.

Berg, G. 1983. *Viral Pollution of the Environment*. Edited by G. Berg. Boca Raton, Fla.: CRC Press.

Bicknell, E.A., D.M. Dziewulski, and L.S. Sturman. 1985. Concentration of Seeded and Naturally Occurring Enteroviruses From Waters of Varying Quality by Hollow Fiber Ultrafiltration. *Water Science and Technology*, 17:47–62.

Brock, T.D. 1983. *Membrane Filtration: A User's Guide and Reference Manual*. Madison, Wis.: Science Technology.

Bukhari, Z., R.M. McCuin, C.R. Fricker, and J.L. Clancy. 1998. Immunomagnetic Separation of *Cryptosporidium parvum* From Source Water Samples of Various Turbidities. *Applied and Environment Microbiology*, 64:4495–4499.

Divizia, M., A.L. Santi, and A. Pana. 1989. Ultrafiltration: An Efficient Second Step for Hepatitis A Virus and Poliovirus Concentration. *Journal of Virological Methods*, 23:55–62.

Dizer, H., J. Durkop, H. Grohman, H. Kopecka, and J.M. Lopez-Pila. 1993. Viruses in Water: Problems, Detection and Control. *Water Science and Technology*, 27:127–133.

Garin, D., F. Fuchs, J.M. Crance, R. Deloince, M. Aymard, and M. Bartoli. 1993. Validation of an Ultrafiltration Process to Concentrate Viruses From Large Volumes of Water. *Environmental Technology*, 14:397–400.

Gerba, C.P. 1996. Pathogens in the Environment. In *Pollution Science*. Edited by I.L. Pepper, C.P. Gerba, and M.L. Brusseau. New York: Academic Press.

Gerba, C.P., C.H. Stagg, C. Wallis, and J.L. Melnick. 1978. Concentration of Enteroviruses From Large Columns of Tap Water, Treated Sewage and Seawater. *Applied and Environment Microbiology*, 35:540–548.

Goyal, S.M., and C.P. Gerba. 1982. Concentration of Viruses From Water by Membrane Filters. In *Methods in Environmental Virology*. Edited by C.P. Gerba and S.M. Goyal. New York: Marcel Dekker.

Graff, J., J. Ticehurst, and B. Flehmig. 1993. Detection of Hepatitis A Virus in Sewage Sludge by Antigen Capture Polymerase Chain Reaction. *Applied and Environment Microbiology*, 59:3165–3170.

Green, J.H., and M. Tylla. 1998. A Comparison of Ultrafiltration on Various River Waters. *Desalination*, 119:79–84.

Guttman-Bass, N., and R. Armon. 1983. Concentration of Simian Rotavirus SA-11 From Tap Water by Membrane Filtration and Organic Flocculation. *Applied and Environment Microbiology*, 45:850–855.

Guttman-Bass, N., and J. Catalano-Sherman. 1985. Effects of Humic Materials on Virus Recovery From Water. *Applied and Environment Microbiology*, 49:1260–1264.

Hill, V.R., A.L. Polaczyk, H. Donghyun, N. Jothikumar, T.L. Cromeans, J.M. Roberts, and J.E. Amburgey. 2005. Development of a Rapid Method for Simultaneous Recovery of Diverse Microbes in Drinking Water by Ultrafiltration With Sodium Polyphosphate and Surfactants. *Applied and Environment Microbiology*, 71:6878–6884.

Jansons, J., and M. Bucens. 1986. Virus Detection in Water by Ultrafiltration. *Water Research*, 12:1603–1606.

Jothikumar, N., D.O. Cliver, and T.W. Mariam. 1998. Immunomagnatic Capture PCR for Rapid Concentration and Detection of Hepatitis A Virus From Environmental Samples. *Applied and Environment Microbiology*, 64:504–508.

Juliano, J., and M.D. Sobsey. 1998. Simultaneous Concentration of *Cryptosporidium*, Bacteria, and Viruses by Hollow-Fiber Ultrafiltration. In *Proc. of the Water Quality Technology Conference 1998* [CD-ROM]. Denver, Colo.: AWWA.

Kfir, R., C. Hilner, M. Du Preez, and B. Bateman. 1995. Studies Evaluating the Applicability of Utilizing the Same Concentration Techniques for the Detection of Protozoan Parasites and Viruses in Water. *Water Science and Technology*, 31:417–423.

Knabel, S.J., H.W. Walker, and P.A. Hartman. 1991. Inhibition of *Aspergillus flavus* and Selected Gram-Positive Bacteria by Chelation of Essential Metal Cations by Polyphosphates. *Journal of Food Protection*, 54:360–365.

Kuhn, R.C., and K.H. Oshima. 2001. Evaluation and Optimization of a Reusable Hollow Fiber Ultrafilter as a First Step in Concentrating *Cryptosporidium* Oocysts From Water. *Water Reseach*, 35:2779–2783.

Kuhn, R.C., and K.H. Oshima. 2002. Hollow-Fiber Ultrafiltration of *Cryptosporidium parvum* Oocysts From a Wide Variety of 10-L Surface Water Samples. *Canadian Journal of Microbiology*, 48:542–549.

Lantz, P.-G., M. Matsson, T. Wadström, and P. Rådström. 1997. Removal of PCR Inhibitors From Human Fecal Samples Through the Use of an Aqueous Two-Phase System for Sample Preparation Prior to PCR. *Journal of Microbiological Methods*, 28:159–167.

Lepesteur, M., S. Blasdall, and N.J. Ashbolt. 2003. Particle Dispersion for Further *Cryptosporidium* and *Giardia* Detection by Flow Cytometry. *Letters in Applied Microbiology*, 37:218–229.

Melnick, J.L., R. Safferman, V.C. Rao, S. Goyal, G. Berg, D.R. Dahling, B.A. Wright, E. Akin, R. Stetler, C. Sorber, M.D. Sobsey, R. Moore, A.L. Lewis, and F.M. Wellings. 1984. Round Robin Investigation of Methods for the Recovery of Poliovirus From Drinking Water. *Applied and Environment Microbiology*, 47:144–150.

Morales-Morales, H.A., G. Vidal, J. Olszewski, C.M. Rock, D. Dasgupta, K.H. Oshima, and G.B. Smith. 2003. Optimization of a Reusable Hollow-Fiber Ultrafilter for Simultaneous Concentration of Enteric Bacteria, Protozoa, and Viruses From Water. *Applied and Environment Microbiology*, 69:4098–4102.

Olszewski, J., L.J. Winona, and K.H. Oshima. 2001. Hollow Fiber Ultrafiltration to Concentrate Viruses From Environmental Waters. In *Proc. of the Water Quality Technology Conference 2001* [CD-ROM]. Denver, Colo.: AWWA.

Olszewski, J., L.J. Winona, and K.H. Oshima. 2005. Comparison of 2 Ultrafiltration Systems for the Concentration of Seeded Viruses From Environmental Waters. *Canadian Journal of Microbiology*, 51:295–303.

Oshima, K.H. 2001. *Efficient and Predictable Recovery of Viruses and* Cryptosporidium *parvum Oocysts From Water by Ultrafiltration Systems.* Technical Completion Report 01-4-23949. Las Cruces, N.M.: New Mexico Water Research Resources Institute, New Mexico State University Department of Biology.

Oshima, K.H., T.T. Evans-Strickfaden, A.K. Highsmith, and E.W. Ades. 1995. The Removal of Phages T1 and PP7 and Poliovirus From Fluids With Hollow-Fiber Ultrafilters With Molecular Weight Cut-Offs of 50,000, 13,000, and 6,000. *Canadian Journal of Mirobiology*, 41:316–322.

Papo, A., L. Piani, and R. Ricceri. 2002. Sodium Tripolyphosphate and Polyphosphate as Dispersing Agents for Kaolin Suspensions: Rheological Characterization. *Colloids and Surfaces A: Physicochemical and Engineering Aspects,* 201:219–230.

Payment, P., A. Berube, D. Perreault, R. Armon, and M. Trudel. 1989. Concentration of *Giardia lamblia* cysts, *Legionella pneumophila*, *Clostridium perfringens*, Human Enteric Viruses, and Coliphages From Large Volumes of Drinking Water, Using a Single Filtration. *Canadian Journal of Microbiology*, 35:932–935.

Rose, J.B., S.N. Singh, C.P. Gerba, and L.M. Kelley. 1984. Comparison of Microporous Filters for Concentration of Viruses From Wastewater. *Applied and Environment Microbiology*, 47:989–992.

Schwab, K.J., R. De Leon, and M.D. Sobsey. 1995. Concentration and Purification of Beef Extract Mock From Water Samples for the Detection of Enteroviruses, Hepatitis A Virus and Norwalk Virus by Reverse Transcriptase–PCR. *Applied and Environment Microbiology*, 61:531–537.

Sharma, M.M., Y.I. Chang, and T.F. Yen. 1985. Reversible and Irreversible Surface Charge Modification of Bacteria for Facilitating Transport Through Porous Media. *Colloids and Surfaces,* 16:193–206.

Shields, P.A., and S.R. Farrah. 1983. Influence of Salts on Electrostatic Interactions Between Poliovirus and Membrane Filters. *Applied and Environment Microbiology*, 45:526–531.

Shields, P.A., T.F. Ling, V. Tjatha, D.O. Shah, and S.R. Farrah. 1986. Comparison of Positively Charged Membrane Filter Methods. *Water Research*, 20:145–151.

Simmons, O., J.J. Juliano, C.D. Heaney, D.S. Francy, F.W. Schaefer III, and M.D. Sobsey. 1999. Comparison of Filtration Methods for Primary Recovery of *Cryptosporidium parvum* From Water. In *Proc. of the International Symposium on Waterborne Pathogens.* [CD-ROM.] Denver, Colo.: AWWA.

Simmons, O.D., M.D. Sobsey, C.D. Heaney, F.W. Schaefer III, and D.S. Francy. 2001. Concentration and Detection of *Cryptosporidium* Oocysts in Surface Water Samples by Method 1622 Using Ultrafiltration and Capsule Filtration. *Applied and Environment Microbiology*, 67:1123–1127.

Sobsey, M.D. 1976. Methods for Detecting Enteric Viruses in Water and Wastewater. In *Viruses in Water*. Edited by G. Berg, H.L. Bodily, E.H. Lennette, J.L. Melnick, and T.G. Metcalf. Washington, D.C.: APHA.

Sobsey, M.D., and J.S. Glass. 1984. Influence of Water Quality on Enteric Virus Concentration by Microporous Filter Methods. *Applied and Environment Microbiology*, 47:956–960.

Sobsey, M.D., and A.R. Hickey. 1985. Effects of Humic and Fulvic Acids on Poliovirus Concentration From Water by Microporous Filtration. *Applied and Environment Microbiology*, 49:259–264.

Straub, T.M., and D.P. Chandler. 2003. Towards a Unified System for Detecting Waterborne Pathogens (review article). *Journal of Microbiological Methods*, 53:185–197.

Suttle, C.A., A.M. Chan, and M.T. Cottrell. 1991. Use of Ultrafiltration to Isolate Viruses From Seawater Which Are Pathogens of Marine Phytoplankton. *Applied and Environment Microbiology*, 57:721–726.

USEPA (United States Environmental Protection Agency). 2001a. Method 1602: Male-Specific (F+) and Somatic Coliphage in Water by Single Agar Layer (SAL) Procedure. EPA821-R-01-029. Washington, D.C.: USEPA, Office of Water.

USEPA (United States Environmental Protection Agency). 2001b. Method 1623: *Cryptosporidium* and *Giardia* in Water by Filtration/IMS/IFA. EPA 821-R-01-025. Washington, D.C.: USEPA, Office of Water.

USEPA (United States Environmental Protection Agency). 2002a. Method 1106.1: Enterococci in Water by Membrane Filtration Using Membrane-Enterococcus-Esculin Iron Agar (mE-EIA). EPA 821-R-02-021. Washington, D.C.: USEPA, Office of Water.

USEPA (United States Environmental Protection Agency). 2002b. Method 1603: *Escherichia coli* (*E. coli*) in Water by Membrane Filtration Using Modified Membrane–Thermotolerant *Escherichia Coli* Agar (Modified mTEC). EPA 821-R-02-023. Washington, D.C.: USEPA, Office of Water.

USEPA (United States Environmental Protection Agency). 2002c. Method 1604: Total Coliforms and *Escherichia coli* in Water by Membrane Filtration Using a Simultaneous Detection Technique (MI Medium). EPA 821-R-02-024. Washington, D.C.: USEPA, Office of Water.

Venczel, L.V., M. Arrowood, M. Hurd, and M.D. Sobsey. 1997. Inactivation of *Cryptosporidium parvum* Oocysts and *Clostridium perfringens* Spores by a Mixed-Oxidant Disinfectant. *Applied and Environment Microbiology*, 63:1598–1601.

Vincent, R.H., A.L. Polaczyk, D. Hahn, J. Narayanan, T.L. Cromeans, J.M. Roberts, and J.E. Amburgey. 2005. Development of a Rapid Method for Simultaneously Recovering Microbes in Drinking Water Using Ultrafiltration With Sodium Polyphosphate and Surfactants. *Applied and Environment Microbiology*, 71:6878–6884.

Wilson, I.G. 1997. Inhibition and Facilitation of Nucleic Acid Amplification. *Applied and Environment Microbiology*, 63:3741–3751.

Winona, L.J., A.W. Omani, J. Olszewski, J.B. Nuzzo, and K.H. Oshima. 2001. Efficient and Predictable Recovery of Viruses From Water by Small Scale Ultrafiltration Systems. *Canadian Journal of Microbiology*, 47:1033–1041.

ABBREVIATIONS

ACS	American Chemical Society
APHA	American Public Health Association
ATCC	American Type Culture Collection
AWWA	American Water Works Association
AwwaRF	Awwa Research Foundation
BSA	bovine serum albumen
°C	degrees Celsius
cc	cubic centimeter
cfu	colony-forming unit
cm	centimeter
cm^2	square centimeter
Da	Dalton
DI	deionized
DMEM-10	Dulbeccos minimum essential media with 10% fetal bovine or calf serum
DPD	N,N-diethyl-p-phenylenediamine
ft^2	square foot
g	gravity
g	gram
GSM	generic spore media
hr	hour
IFA	immunofluorescent antibody
IMS	immunomagnetic separation
in.	inch
kPa	kilopascal
L	liter
m^2	square meter
M	molar
max.	maximum
MEM	minimum essential media/medium
mg	milligram
mg/L	milligrams per liter
min	minute
mL	milliliter

mm	millimeter
MWCO	molecular weight cutoff
μm	micrometer
NA	not analyzed
NaPP	sodium polyphosphate
ND	no data
nm	nanometer
OD	optical density
PAN	polyacrylonitrile
PBS	phosphate-buffered saline
pfu	plaque-forming unit
ppm	parts per million
psi	pounds per square inch
RNA	ribonucleic acid
RO	reverse osmosis
rpm	revolutions per minute
SD	standard deviation
SDS	sodium dodecyl sulfate
TNTC	too numerous to count
TSA	trypticase soy agar
UF	ultrafiltration
UPW	ultrapure water
U.S.	United States
USEPA	United States Environmental Protection Agency
v/v	volume per volume
WEF	Water Environment Federation